T0093259

LONDON MATHEMATICAL SOCIETY STUDENT TEXTS

57 Introduction to Banach algebras, operators and harmonic analysis, GARTH DALES *et al*
58 Computational algebraic geometry, HAL SCHENCK
59 Frobenius algebras and 2D topological quantum field theories, JOACHIM KOCK
60 Linear operators and linear systems, JONATHAN R. PARTINGTON
61 An introduction to noncommutative Noetherian rings: Second edition, K. R. GOODEARL & R. B. WARFIELD, JR
62 Topics from one-dimensional dynamics, KAREN M. BRUCKS & HENK BRUIN
63 Singular points of plane curves, C. T. C. WALL
64 A short course on Banach space theory, N. L. CAROTHERS
65 Elements of the representation theory of associative algebras I, IBRAHIM ASSEM, DANIEL SIMSON & ANDRZEJ SKOWROŃSKI
66 An introduction to sieve methods and their applications, ALINA CARMEN COJOCARU & M. RAM MURTY
67 Elliptic functions, J. V. ARMITAGE & W. F. EBERLEIN
68 Hyperbolic geometry from a local viewpoint, LINDA KEEN & NIKOLA LAKIC
69 Lectures on Kähler geometry, ANDREI MOROIANU
70 Dependence logic, JOUKU VÄÄNÄNEN
71 Elements of the representation theory of associative algebras II, DANIEL SIMSON & ANDRZEJ SKOWROŃSKI
72 Elements of the representation theory of associative algebras III, DANIEL SIMSON & ANDRZEJ SKOWROŃSKI
73 Groups, graphs and trees, JOHN MEIER
74 Representation theorems in Hardy spaces, JAVAD MASHREGHI
75 An introduction to the theory of graph spectra, DRAGOŠ CVETKOVIĆ, PETER ROWLINSON & SLOBODAN SIMIĆ
76 Number theory in the spirit of Liouville, KENNETH S. WILLIAMS
77 Lectures on profinite topics in group theory, BENJAMIN KLOPSCH, NIKOLAY NIKOLOV & CHRISTOPHER VOLL
78 Clifford algebras: An introduction, D. J. H. GARLING
79 Introduction to compact Riemann surfaces and dessins d'enfants, ERNESTO GIRONDO & GABINO GONZÁLEZ-DIEZ
80 The Riemann hypothesis for function fields, MACHIEL VAN FRANKENHUIJSEN
81 Number theory, Fourier analysis and geometric discrepancy, GIANCARLO TRAVAGLINI
82 Finite geometry and combinatorial applications, SIMEON BALL
83 The geometry of celestial mechanics, HANSJÖRG GEIGES
84 Random graphs, geometry and asymptotic structure, MICHAEL KRIVELEVICH *et al*
85 Fourier analysis: Part I – Theory, ADRIAN CONSTANTIN
86 Dispersive partial differential equations, M. BURAK ERDOĞAN & NIKOLAOS TZIRAKIS
87 Riemann surfaces and algebraic curves, R. CAVALIERI & E. MILES
88 Groups, languages and automata, DEREK F. HOLT, SARAH REES & CLAAS E. RÖVER
89 Analysis on Polish spaces and an introduction to optimal transportation, D. J. H. GARLING
90 The homotopy theory of $(\infty,1)$-categories, JULIA E. BERGNER
91 The block theory of finite group algebras I, M. LINCKELMANN
92 The block theory of finite group algebras II, M. LINCKELMANN
93 Semigroups of linear operators, D. APPLEBAUM
94 Introduction to approximate groups, M. C. H. TOINTON
95 Representations of finite groups of Lie type (2nd Edition), F. DIGNE & J. MICHEL
96 Tensor products of C^*-algebras and operator spaces, G. PISIER

London Mathematical Society Student Texts 94

Introduction to Approximate Groups

MATTHEW C. H. TOINTON
University of Cambridge

CAMBRIDGE
UNIVERSITY PRESS

University Printing House, Cambridge CB2 8BS, United Kingdom

One Liberty Plaza, 20th Floor, New York, NY 10006, USA

477 Williamstown Road, Port Melbourne, VIC 3207, Australia

314–321, 3rd Floor, Plot 3, Splendor Forum, Jasola District Centre,
New Delhi – 110025, India

79 Anson Road, #06–04/06, Singapore 079906

Cambridge University Press is part of the University of Cambridge.

It furthers the University's mission by disseminating knowledge in the pursuit of education, learning, and research at the highest international levels of excellence.

www.cambridge.org
Information on this title: www.cambridge.org/9781108470735
DOI: 10.1017/9781108652865

First published 2020

Printed in the United Kingdom by TJ International Ltd. Padstow Cornwall

A catalogue record for this publication is available from the British Library

ISBN 978-1-108-47073-5 Hardback
ISBN 978-1-108-45644-9 Paperback

To Kate and Amelia, with love

Contents

Preface

Mathematicians are used to the notion of a subgroup of a group G as a subset containing the identity that is closed under taking products and inverses. However, it turns out that there are also circumstances in which we encounter subsets that are merely 'approximately closed'. Such sets arise, for example, in the construction of *expander graphs*, which are important in theoretical computer science, or in the study of *polynomial growth* in geometric group theory, which in turn has links to random walks and differential geometry. There are also numerous other examples.

A priori, there are a number of different ways of defining approximate closure. The notion that tends to arise in applications is that of *small doubling*, which we introduce in Definition 1.1.1. A more sophisticated and in some ways more tractable notion is the one that gives its name to this book and to the theory: that of an *approximate subgroup*. We introduce this in Definition 1.1.2. As we shall see, the two notions are intimately linked, and in some sense ultimately interchangeable. The aim of this book is to motivate and develop these notions in detail, with a view to leaving the reader in a position to understand and add to their growing literature, as well as to apply them elsewhere.

It turns out that the name *approximate subgroup* is justified by more than its origins as a notion of approximate closure. Indeed, we shall see in Sections 2.3 and 2.6 that many of the properties of approximate subgroups can be viewed as approximate versions of properties of 'exact' subgroups (as we will occasionally call genuine subgroups to emphasise their relationship to approximate subgroups). Understanding which properties of exact subgroups persist when we pass to approximate subgroups, and to what extent they persist, is an important part of the theory.

A striking feature of the theory of approximate groups is the range of fields that it uses and can be applied to. In this book alone we make heavy use of tools and ideas from combinatorics, convex geometry, group theory, representation theory and harmonic analysis, as well as touching on notions from probability. There are also substantial results in the literature on approximate groups that rest on algebraic group theory and model theory, although in order to keep the book reasonably focused we do not present these arguments here, instead directing the reader to the relevant references. Moreover, in addition to the applications to expander graphs and polynomial growth mentioned above, approximate groups have been applied to sieve theory, additive combinatorics, differential geometry and random walks, to name a few.

The applications of approximate groups are too numerous and diverse to present comprehensively in this book. However, since the applicability of approximate groups is one of their great selling points, it would seem remiss not to include at least some of them. We shall therefore go into some detail on certain applications to polynomial growth, and to geometric group theory more generally, in Chapter 11. The choice to present these particular applications rather than any others reflects my own interests as much as anything. For a taste of some other applications the reader may care to read Green's survey article [33].

One final important feature of the theory of approximate groups is that it is to some extent still being worked out. The elementary theory seems to be rather settled, but essentially all of the deeper results appearing in this book still have room for improvement. The book should therefore be thought of as giving a snapshot of the current state of an active research topic, rather than being a definitive description of its final form. Indeed, one of the main aims of the book is to equip the next generation of researchers with the tools and techniques that will enable them to take the field further. I hope that in a few years' time I will have the opportunity to write a revised version incorporating improvements to the theory made by the readers of the present version.

Acknowledgements

I am grateful to Emmanuel Breuillard for having encouraged the development of this book, and more generally for having been generous with his time, advice and support for much of my mathematical career.

I am also grateful to David Tranah at Cambridge University Press for his enthusiasm and support for this project from a very early stage.

I thank Ben Green for having introduced me to both approximate groups and growth in groups. Parts of Chapters 2–4 and Chapter 11 are inspired by two Part III courses he gave whilst I was a student in Cambridge, and I benefited from having access to some unpublished notes of his on growth.

I thank Emmanuel Breuillard, Pierre de la Harpe, Clare Dennison, Ben Green, Barnabás Janzer, Kate Marshall, Stuart Martin, Terence Tao, Romain Tessera, Jaspal Thandi and Matthew Wales for various helpful comments, corrections and discussions, as well as the anonymous referees whose comments improved the papers [72, 73] on which Chapters 6 and 8 are based. In 2019 I gave a Cambridge Part III course based on parts of this book, and I am also grateful to everyone who attended that course (some of whom are named above) for their comments, discussion and enthusiasm.

I am especially grateful to Clare Dennison at Cambridge University Press for her considerable support throughout the writing and production process, and Richard Hutchinson, the book's copyeditor, for a careful and thorough reading of the manuscript that improved its readability and considerably reduced the number of errors.

At different stages of the writing of this book I was based at Homerton College (supported by a Junior Research Fellowship), the Université de Neuchâtel (supported by grant FN 200021_163417/1 of the Swiss National Fund for scientific research), and Pembroke College (where I am the Stokes Research Fellow). I am grateful to all of these institutions for their generosity, hospitality and flexibility.

Matthew Tointon

Pembroke College, Cambridge
September 2019

1
Introduction

1.1 Introduction

As we described in the preface, the theory of approximate groups can be thought of as describing those subsets of groups that are 'approximately closed'. We start by presenting a preliminary notion of approximate closure: *small doubling*. Given two sets A, B inside a group G we define their *product set* $AB = \{ab : a \in A, b \in B\}$. We also write A^{-1} for the set of inverses of elements of A, and write A^n and A^{-n} to denote the *iterated product sets* defined recursively by $A^0 = \{1\}$, $A^n = AA^{n-1}$ and $A^{-n} = (A^{-1})^n$. The study of product sets began in the setting of abelian groups, where one traditionally uses additive notation. Thus, if G is abelian we define the *sum set* $A + B = \{a + b : a \in A, b \in B\}$ and the *difference set* $A - B = \{a - b : a \in A, b \in B\}$. We also write $-A$ for the set of inverses of elements of A, and write nA and $-nA$ in place of A^n and A^{-n}, respectively.

To say that a finite set A is closed under taking products is then to say that $A^2 = A$. One way to define 'approximate' closure is to say that A^2 is not too much larger than A. To get a feel for what this might mean in practice, let us consider for a moment what might be thought of as 'extremal' or 'typical' for the size of A^2. It is not difficult to see what the extremal possibilities for $|A^2|$ are in terms of $|A|$: it is clear that $|A| \leq |A^2| \leq |A|^2$, and in general neither bound can be improved. Indeed, if A is a finite subgroup of G then $|A^2| = |A|$, while if G is the free group generated by A then $|A^2| = |A|^2$.

It turns out that the quadratic upper bound on the size of A^2 is in fact typical in some sense. For example, we show in Section 2.1 that if A is a set of size k chosen uniformly at random from an interval $\{1, \ldots, n\} \subset \mathbb{Z}$ with n much larger than k then $\mathbb{E}[|A^2|]$ is close to $k^2/2$. This suggests that a 'generic' set A should have $|A^2|$ comparable to $|A|^2$, and so it is sets for which

$$|A^2| = o(|A|^2) \tag{1.1.1}$$

that we should view as being 'exceptional'.

The theory of approximate groups is essentially concerned with the extreme case of (1.1.1) in which $|A^2|$ is *linear* in $|A|$, in the sense that

$$|A^2| \leq K|A| \tag{1.1.2}$$

for some fixed $K \geq 1$. Since (1.1.2) represents 'non-random' behaviour, we can expect such sets to exhibit a certain amount of 'structure'. One of the principal aims of approximate-group theory, and of this book, is to describe this structure in as much detail as possible.

Of course, one type of structure satisfying (1.1.2) is a finite subgroup, for which we may even take $K = 1$. Another trivial example is if A itself has size at most K. Let us reassure ourselves, though, that the theory of sets satisfying (1.1.2) is more general than just the theory of finite subgroups and 'small' sets. Indeed, it is easy to see that the set $A = \{-n, \dots, n\} \subset \mathbb{Z}$ satisfies $|A + A| \leq 2|A|$, and so the group $\mathbb{Z}$ contains arbitrarily large finite sets of small doubling, even though it contains no non-trivial subgroups. We will develop and generalise this example in Chapter 3.

Since it is the key property that we will be investigating, we now give a name to those sets satisfying (1.1.2).

Definition 1.1.1 (small doubling) Given a finite subset A of a group we call the quantity $|A|^2/|A|$ the *doubling constant* of A. If the doubling constant of A is at most a given constant K then we often say simply that A is a *set of doubling at most K*, or even merely a *set of small doubling*.

As we shall explain in some detail in Chapter 2, in some contexts, and particularly in the case of non-abelian groups, it is convenient for technical reasons to replace Definition 1.1.1 with a slightly stronger definition, due to Tao, which gives its name both to this book and to the theory.

Definition 1.1.2 (approximate group) A subset A of a group G is said to be a *K-approximate subgroup of G*, or simply a *K-approximate group*, if $A^{-1} = A$ and $1 \in A$, and if there exists $X \subset G$ with $|X| \leq K$ such that $A^2 \subset XA$.

Note in particular that a finite K-approximate group has doubling at most K. The conditions $A^{-1} = A$ and $1 \in A$ are largely for notational convenience. On the one hand, assuming that $A^{-1} = A$ avoids the need to distinguish between positive and negative iterated products, allowing us to replace an untidy-looking expression such as $A^2A^{-3}AA^{-1}A^3 \cup A^{-4}A^3A^{-1}$ with the more succinct A^{10}, for example. On the other hand, assuming that $1 \in A$ means that we have the nesting $A \subset A^2 \subset A^3 \subset \cdots$, which is also convenient at times. The existence of $X \subset G$ with $|X| \leq K$ such that $A^2 \subset XA$ is more serious, however. Indeed, we shall see in Chapter 2 that one can construct sets of bounded doubling that fail to be K-approximate groups for arbitrarily large K, so being a finite approximate group is strictly stronger than having small doubling. However, when introducing the definition of approximate groups Tao showed that the study of sets of small doubling nonetheless essentially reduces to the study of approximate groups in a certain precise way; in Theorem 2.5.6 we present a strengthening of this reduction that follows from work of Petridis.

One specific advantage of Definition 1.1.2 over Definition 1.1.1 that is worth emphasising at this point is that it applies without modification to infinite subsets of groups. Indeed, there has recently begun to emerge a theory of infinite approximate groups in certain particular contexts (see [6], for example). Nonetheless, the theory of finite approximate groups is far more developed than the theory of infinite approximate groups, and is the focus of this book.

In Chapter 2 we motivate and develop Definitions 1.1.1 and 1.1.2 in more detail, in particular deriving some of their elementary properties. In Chapter 3 we look in detail at some specific examples of sets of small doubling and approximate groups. In the largest part of the book, comprising Chapters 4–10, we prove a number of results describing the structure of approximate subgroups in various classes of group. Finally, in Chapter 11 we present some applications of approximate groups to geometric group theory.

1.2 Historical Discussion

In this section we very briefly present the historical context of the material of this book. We stress that this is designed to give the reader an overall feel for the development of the theory, rather than to be a comprehensive history.

Much of the early progress on classifying sets of small doubling focused on abelian groups. The theory was initiated in the 1960s by Freiman [26], who in particular gave an essentially complete classification of sets of small doubling in the integers. The theory was subsequently developed considerably by Ruzsa, who amongst other things gave a simpler proof of Freiman's theorem [55]. Ruzsa's work was brought to the attention of a wider audience when Gowers [30, 31] applied it in his celebrated proof of a theorem of Szemerédi [61] concerning arithmetic progressions in dense sets of integers. In the mid 2000s, Green and Ruzsa [35] generalised Freiman's theorem to arbitrary abelian groups; we present their result in Chapter 4.

Another important early result on abelian groups was the so-called *sum–product theorem* of Bourgain, Katz and Tao [10]. This roughly states that a subset of $\mathbb{F}_p$ cannot simultaneously have small additive doubling and small multiplicative doubling, unless it is either very small or already almost all of $\mathbb{F}_p$. One of the tools used in the proof was a result from Gowers's work on Szemerédi's theorem, refining work of Balog and Szemerédi and now often known as the *Balog–Szemerédi–Gowers theorem*. We introduce this briefly as Theorem 2.1.5. We discuss sum–product theorems further in Section 9.2.

At around the same time as Green and Ruzsa's generalisation of Freiman's theorem, efforts began in earnest to generalise these concepts and results to non-abelian groups. Some of the first work in this direction was by Helfgott [39], who showed that a generating subset of A of $SL_2(\mathbb{Z}/p\mathbb{Z})$ does not even satisfy the weaker version $|A| \leq c|A|^{1+\varepsilon}$ of (1.1.1), unless it is already close to the whole of $SL_2(\mathbb{Z}/p\mathbb{Z})$. Amongst the tools used by Helfgott were aspects of Ruzsa's theory, the Bourgain–Katz–Tao sum–product theorem, and the Balog–Szemerédi–Gowers theorem. Helfgott's result is of particular interest because of its use by Bourgain and Gamburd [9] to construct so-called *expander graphs*, one of the most celebrated applications of the theory.

The first systematic account of the elementary theory of sets of small doubling in non-abelian groups was Tao's foundational work [62]. This work also introduced the notion of approximate groups and proved their essential equivalence to small doubling (although as we note in Remark 2.4.8 the definition of approximate groups was to some extent anticipated by Green and Ruzsa). We present much of this material in Chapter 2.

After Tao's work there were a number of papers in fairly quick succession proving Freiman- or Helfgott-type results for various non-abelian groups, such as soluble groups (Tao [65]), free groups (Razborov [50] and Safin [57]), torsion-free nilpotent groups (Breuillard–Green [12]), and various linear groups (Breuillard–Green [13, 14] and Gill–Helfgott [28]). We present some of these results in this book; for example, in Chapter 6 we generalise the result of [12] to arbitrary nilpotent groups, and in Chapter 9 we present the result of [13].

There has also been much subsequent work on generalising Helfgott's work and its applications to expansion, notably by Pyber and Szabó [49] and Breuillard, Green and Tao [16]. We describe this briefly in an appendix to Chapter 11, but Tao's book [66] already gives an excellent and comprehensive account of this work, so we refer the interested reader to that source for the details rather than repeating them here.

It turns out that many of the results discussed above are somewhat reminiscent of a phenomenon seen in the related context of *polynomial growth.* A subset A of a group exhibits *polynomial growth* if there exists a polynomial p such that $|A^n| \leq p(n)$ for all $n \in \mathbb{N}$. One slightly imprecise but intuitively useful way of comparing this to Definition 1.1.1 is that whilst Definition 1.1.1 says that A 'grows slowly' when it is multiplied by itself once, polynomial growth means that A 'grows slowly' when it is multiplied by itself any number of times. Moreover, a famous theorem of Gromov describing the structure of sets of polynomial growth exploits the easily checked fact that if A is such a set then there are infinitely many n for which A^n has small doubling. Gromov's theorem states that if A has polynomial growth then the group generated by A has a *nilpotent* subgroup of finite index (for readers unfamiliar with nilpotence, we give a detailed introduction in Chapter 5). As we will see in this book, many of the results listed above show that sets of small doubling in the groups under consideration also have a significant amount of nilpotent structure in some sense.

Helfgott and Lindenstrauss conjectured that these similarities between Gromov's theorem and results on sets of small doubling were not coincidental, and that in fact an arbitrary approximate subgroup should have a large amount of nilpotent structure in a precise sense. This was finally proved in 2011 by Breuillard, Green and Tao [18]. Their result, which we state in Chapter 7, essentially describes the structure of an arbitrary approximate group. It also leads to a refinement of Gromov's theorem, and in turn to various other applications to geometric group theory, some of which we describe in Chapter 11.

We end this historical note by emphasising that the history of approximate groups is still being written. In particular, the reader should not interpret the existence of the Breuillard–Green–Tao result as meaning that the theory is complete. Indeed, whilst that result is very general, as we explain in Chapter 7 its conclusion is rather imprecise in a particular, quantitative sense. Indeed, even the optimal classification of sets of small doubling in abelian groups is not yet known, and, as we said in the preface, essentially all of the results of this type that we present in this book have room for improvement.

1.3 Bounds and Asymptotic Notation

The larger K is in Definitions 1.1.1 and 1.1.2, the weaker they become. We can therefore expect that the structure of a set satisfying Definition 1.1.1 or 1.1.2 that we are able to obtain should become 'rougher' as K increases. A big part of the results we present will be to quantify this increased 'roughness'. For example, in Theorem 2.2.1 we show that if A is a finite subset of a group satisfying Definition 1.1.1 with $K < \frac{3}{2}$ then there exists a subgroup H such that A lies in a coset of H and $|A|/|H| \geq 1/K$. Thus, A is a 'large' proportion of a coset of a subgroup, and the meaning of 'large' depends on K in a precise, quantified way.

At times, however, the precise expression we obtain in terms of K is less important than the overall form it takes. For example, if one result says that A is a subset of a certain structure H with $|A|/|H| \geq \exp(-15K^3 + \log K)$, and another says the same thing but with $|A|/|H| \geq K^{-17}/100$, the fact that the first bound is exponential but the second is merely polynomial is far more important than the precise values of the constants or exponents in these expressions. In this specific setting, one might reasonably choose simply to say that there exist absolute constants $c, C > 0$ such that $|A|/|H| \geq cK^{-C}$ in the case of the first result or $|A|/|H| \geq \exp(-CK^C)$ in the case of the second (to say that a constant is *absolute* here means that it does not depend in any way on A or K). We therefore deploy the some standard shorthand notation to abbreviate bounds such as these in a way that emphasises the important 'shape' of the bound without the distraction of inconsequential constants and exponents, as follows.

We follow the standard convention that if X, Y are real, variable quantities then $X \ll Y$ and $Y \gg X$ each mean that there exists a constant $C > 0$ such that X is always at most CY. Thus, for example, one may write $10n^2 \ll n^3$ for $n \in \mathbb{N}$ because, for example, $10n^2 \leq 10n^3$ for every

$n \in \mathbb{N}$. We call C the constant *implicit in* or *implied by* the $\ll$ or $\gg$ notation.

The notation $O(Y)$ denotes a quantity that is at most a certain constant multiple of Y, while $\Omega(X)$ denotes a quantity that is at least a certain positive constant multiple of X. Thus, for example, we write $A \subset B^{O(1)}$ to mean that there exists a constant C and a number $m \le C$ such that $A \subset B^m$, or say that a subgroup H is of index $O(m)$ in G to mean that there exists a constant C such that $[G : H] \le Cm$. Technically the O and Ω notation could be used to replace the $\ll$ and $\gg$ notation, but we tend to opt for $\ll$ and $\gg$ where possible.

In the $\ll, \gg, O, \Omega$ notation, if the constant in question depends on some other variable z then we indicate this with a subscript, for example $X \ll_z Y$ or $O_z(Y)$.

The reader may find it a useful exercise to check that he or she has understood the above notation by verifying that

$$K^K \le \exp(K^{O(1)})$$

for $K > 0$, a bound that we use frequently in the book without explicit mention.

Despite the importance of the bounds in many of the theorems we prove, in a number of cases where we have the option to simplify an argument at the expense of making the bounds worse we opt to do so, a trade-off one would usually not make in a research paper, but which suits the pedagogical aims of this book. Nonetheless, we always provide references to arguments giving the best bounds the author is aware of.

1.4 General Notation

We assume familiarity with the basic concepts, definitions and results from group theory that can be found in a book such as Hall [38] or Robinson [51]. In particular, we assume familiarity with the definition of a *free group* as given in [38, §7.1], for example.

Here is a list of specific notation and definitions that we use in this book.

- We write

$$\begin{aligned} \mathbb{N} &= \{1, 2, \ldots\}, \\ \mathbb{N}_0 &= \mathbb{N} \cup \{0\}, \\ [n] &= \{1, \ldots, n\}, \\ [n]_0 &= \{0, \ldots, n\}, \\ [n]^{\pm} &= \{-n, \ldots, n\}. \end{aligned}$$

- We write $\mathbb{C}^\times$ for the set of non-zero complex numbers. Given a prime p, we also write $(\mathbb{Z}/p\mathbb{Z})^\times$ for the set of non-zero elements of $\mathbb{Z}/p\mathbb{Z}$. In each case these sets form groups under the operation of multiplication.
- Given a subset A of an abelian group and $n \in \mathbb{N}$ we define the *dilate* $n \cdot A$ via $n \cdot A = \{na : a \in A\}$.
- Given a subset A of a set X, we write $1_A : X \to \{0, 1\}$ for the indicator function of A defined via

$$1_A(x) = \begin{cases} 1 \text{ if } x \in A \\ 0 \text{ if } x \notin A. \end{cases}$$

  Given a function $f : X \to Y$ into some other set Y, we write $f|_A : A \to Y$ for the restriction of f to A.
- Given $x > 0$ and $c \in \mathbb{R}$ we write $\log^c x$ to mean $(\log x)^c$.
- We use expectation notation to write averages over finite sets. Specifically, given a finite set X and a function $f : X \to \mathbb{C}$ we define

$$\mathbb{E}_{x \in X} f(x) = \frac{1}{|X|} \sum_{x \in X} f(x).$$

- In general we write 1 for the identity element of any group. The main exception to this is that we normally write abelian groups additively, in which case we write 0 for the identity element. When we occasionally use alternative symbols we always state this explicitly.
- Given two sets A, B, we write $A \subset B$ to mean that A is a subset of B. *This allows the possibility that* $A = B$. Given groups G, H, we write $H < G$ to mean that H is a subgroup of G, and $H \lhd G$ to mean that H is a normal subgroup of G, *again in each case allowing the possibility that* $A = B$. To indicate that $A \subset B$ with no possibility of equality we write $A \subsetneq B$.

- We define the *rank* of a finitely generated group to be the size of the smallest or joint-smallest generating set.
- Given a group G and a subset $X \subset G$ we write $\langle X \rangle$ for the subgroup of G generated by X. If X is written with braces then we drop the braces when using the $\langle \cdot \rangle$ notation, for example writing $\langle x_1, \ldots, x_r \rangle$ instead of $\langle \{x_1, \ldots, x_r\} \rangle$.
- Given a group G with a subgroup $H < G$, we write H^G for the *normal closure* of H in G, that is the smallest normal subgroup of G containing H.
- We define the *commutator* $[x, y]$ of two elements in a group G via $[x, y] = x^{-1}y^{-1}xy$. We also indicate conjugation using exponents, defining $x^y = y^{-1}xy$, and more generally $Y^x = \{x^{-1}yx : y \in Y\}$ for a subset $Y \subset G$.
- Let G be a group. Given a subgroup $H < G$, we denote by $N_G(H)$ the normaliser of H in G; thus

  $$N_G(H) = \{g \in G : H^g = H\}.$$

  Given a subset $X \subset G$, we denote by $C_G(X)$ the centraliser of X in G; thus

  $$C_G(H) = \{g \in G : [g, x] = 1 \text{ for every } x \in X\}.$$

  Given, in addition, a normal subgroup $N \lhd G$, we write

  $$C_{G/N}(X) = \{g \in G : [g, x] \subset N \text{ for every } x \in X\}.$$

1.5 Miscellaneous Results

Here are some standard results that are too general to belong in any particular chapter of this book, but useful to be able to refer to. Some proofs are left as exercises, and some are outsourced to standard texts.

Theorem 1.5.1 (fundamental theorem of finitely generated abelian groups [51, 4.2.10]) *Let G be a finitely generated abelian group. Then there exist $r \in \mathbb{N}_0$, primes $p_1, \ldots, p_r$, and $m_0, \ldots, m_r \in \mathbb{N}_0$ such that $G \cong \mathbb{Z}^{m_0} \oplus \mathbb{Z}/p_1^{m_1}\mathbb{Z} \oplus \cdots \oplus \mathbb{Z}/p_r^{m_r}\mathbb{Z}$.*

Recall that a subgroup C of a group G is *characteristic* if $\psi(C) = C$ for every $\psi \in \mathrm{Aut}\,(G)$.

Lemma 1.5.2 *Let $C \lhd N \lhd G$ be groups such that N is normal in G and C is characteristic in N. Then C is normal in G.*

Lemma 1.5.3 *Let $N, H \lhd G$ be normal subgroups of a group G. Then $C_{G/N}(H)$ is also normal in G.*

Given a finite set X and functions $f, g : X \to \mathbb{C}$, translating the Cauchy–Schwarz inequality into the expectation notation described in the preface gives

$$|\mathbb{E}_{x\in X} f(x)g(x)|^2 \leq (\mathbb{E}_{x\in X}|f(x)|^2)(\mathbb{E}_{x\in X}|g(x)|^2). \tag{1.5.1}$$

We also have $|\sum_{x\in X} f(x)|^2 = |\sum_{x\in X} 1_X(x)f(x)|^2$, and so the usual Cauchy–Schwarz inequality gives

$$\left|\sum_{x\in X} f(x)\right|^2 \leq |X| \sum_{x\in X} |f(x)|^2. \tag{1.5.2}$$

Theorem 1.5.4 (Fubini's theorem [5, Theorem 18.3]) *Let $f : \mathbb{R}^d \to \mathbb{R}$ be a measurable function, and suppose that*

$$\int_{x\in\mathbb{R}^d} |F(x)|\, dx < \infty.$$

Then, viewing $\mathbb{R}^d$ as $\mathbb{R}^m \times \mathbb{R}^{d-m}$, we have

$$\begin{aligned}\int_{x\in\mathbb{R}^d} F(x)\, dx &= \int_{x_1\in\mathbb{R}^m} \int_{x_2\in\mathbb{R}^{d-m}} F(x_1, x_2)\, dx_2\, dx_1 \\ &= \int_{x_2\in\mathbb{R}^{d-m}} \int_{x_1\in\mathbb{R}^m} F(x_1, x_2)\, dx_1\, dx_2.\end{aligned}$$

2
Basic Concepts

2.1 Large Doubling of Random Sets of Integers

As we described in Chapter 1, the underlying aim of the theory of approximate groups is to understand the structure of subsets of groups that have small doubling in the sense of Definition 1.1.1. The aims of the present chapter are to motivate this definition, to give examples of some situations where the structure of such sets can be described elementarily, and to develop some general theory.

We start by showing that appropriately defined random subsets of integers tend to have very large doubling, so that Definition 1.1.1 does indeed represent a significant restriction on the set A.

Proposition 2.1.1 *Fix $k \in \mathbb{N}$, and for each $n \geq k$ let $A_{k,n}$ be a subset of $[n]$ of size k chosen uniformly at random. Then*

$$\liminf_{n\to\infty} \mathbb{E}[|A_{k,n} + A_{k,n}|] \geq \frac{k^2}{2+\frac{1}{k}}.$$

Remark 2.1.2 Write $D(k) = \max\{|A + A| : A \subset \mathbb{Z}, |A| = k\}$. We will see in Exercise 2.2 that Proposition 2.1.1 implies

$$\frac{\liminf_{n\to\infty} \mathbb{E}[|A_{k,n} + A_{k,n}|]}{D(k)} \to 1$$

as $k \to \infty$. Thus Proposition 2.1.1 shows that 'random subsets of $[n]$ have doubling asymptotically as large as possible'.

Our approach is to show that there are few quadruples $(a, b, c, d) \in A \times A \times A \times A$ satisfying

$$a + b = c + d.$$

We call such quadruples *additive quadruples*. To that end, we define the *additive energy* $E(A)$ of a set A in an abelian group to be the total number of additive quadruples in $A \times A \times A \times A$; thus

$$E(A) = |\{(a, b, c, d) \in A \times A \times A \times A : a + b = c + d\}|.$$

One can define analogously the *multiplicative energy* of a finite subset of a non-abelian group; we refer the reader to [62, §4] or [68, §2.3] for a general discussion of these notions.

Note that we have the trivial upper bound $E(A) \leq |A|^3$, since for every triple $(a, b, c) \in A \times A \times A$ there is at most one element $d \in A$ such that $a + b = c + d$. Moreover, this bound is attained in the case where A is a finite group. In order to make the additive energy less dependent on the size of A, we define the *normalised additive energy* $\omega(A)$ of A by

$$\omega(A) = \frac{E(A)}{|A|^3}.$$

Lemma 2.1.3 *For every finite set A we have*

$$\frac{|A + A|}{|A|} \geq \frac{1}{\omega(A)}.$$

Proof For each $s \in A + A$ write $r(s) = |\{(a, b) \in A \times A : a + b = s\}|$, and note that

$$\sum_{s \in A+A} r(s) = |A|^2 \qquad \text{and} \qquad \sum_{s \in A+A} r(s)^2 = E(A).$$

The Cauchy–Schwarz inequality (1.5.2) therefore gives

$$|A|^4 = \left(\sum_{s \in A+A} r(s) \right)^2 \leq |A + A| \sum_{s \in A+A} r(s)^2 = |A + A| \cdot E(A),$$

which proves the lemma. □

Lemma 2.1.4 *Fix $k \in \mathbb{N}$. Then*

$$\mathbb{E}[\,\omega(A_{k,n})\,] \leq \frac{2k + 1 + o(1)}{k^2}$$

as $n \to \infty$.

Proof We abbreviate $A = A_{k,n}$ throughout. Writing Q for the set of additive quadruples in $[n]^4$ and $p(q) = \mathbb{P}[\,q \in A \times A \times A \times A\,]$ for $q \in Q$, note that

$$\mathbb{E}[\,E(A)\,] = \sum_{q \in Q} p(q).$$

Define subsets $Q_1, Q_2, Q_3, Q_4, Q_5, Q_6$ of Q by

$$\begin{aligned} Q_1 &= \{(a,b,c,d) \in Q : a,b,c,d \text{ distinct}\} \\ Q_2 &= \{(a,a,b,c) \in Q : a,b,c \text{ distinct}\} \\ Q_3 &= \{(a,b,c,c) \in Q : a,b,c \text{ distinct}\} \\ Q_4 &= \{(a,b,a,b) \in Q : a,b \text{ distinct}\} \\ Q_5 &= \{(a,b,b,a) \in Q : a,b \text{ distinct}\} \\ Q_6 &= \{(a,a,a,a) \in Q\}, \end{aligned}$$

so that

$$\mathbb{E}[\,E(A)\,] = \sum_{i=1}^{6} \sum_{q \in Q_i} p(q). \tag{2.1.1}$$

For $q \in Q_1$ we have $p(q) \le (k/n)^4$; for q in Q_2 or Q_3 we have $p(q) \le (k/n)^3$; for q in Q_4 or Q_5 we have $p(q) \le (k/n)^2$; and for q in Q_6 we have $p(q) = k/n$. On the other hand, we have

$$\begin{aligned} |Q_1| &\le n^3 \\ |Q_2| &\le n^2 \\ |Q_3| &\le n^2 \\ |Q_4| &= n^2 - n \\ |Q_5| &= n^2 - n \\ |Q_6| &= n. \end{aligned}$$

It follows that

$$\begin{aligned} \textstyle\sum_{q \in Q_1} p(q) &\le k^4/n \\ \textstyle\sum_{q \in Q_2} p(q) &\le k^3/n \\ \textstyle\sum_{q \in Q_3} p(q) &\le k^3/n \\ \textstyle\sum_{q \in Q_4} p(q) &\le k^2 \\ \textstyle\sum_{q \in Q_5} p(q) &\le k^2 \\ \textstyle\sum_{q \in Q_6} p(q) &= k, \end{aligned}$$

and so (2.1.1) implies that

$$\mathbb{E}[\,E(A)\,] \le \left(\frac{k^4 + 2k^3}{n} + 2k^2 + k \right).$$

In particular, this means that

$$\mathbb{E}[\,\omega(A)\,] \leq \frac{2k+1+o(1)}{k^2}$$

as $n \to \infty$, as required. □

Proof of Proposition 2.1.1 Abbreviating $A = A_{k,n}$, we have

$$\begin{aligned}
\mathbb{E}[|A+A|] &\geq k\mathbb{E}\left(\frac{1}{\omega(A)}\right) && \text{(by Lemma 2.1.3)} \\
&\geq \frac{k}{\mathbb{E}[\,\omega(A)\,]} && \text{(by Jensen's inequality)} \\
&\geq \frac{k^2}{2+\frac{1}{k}+o(1)} && \text{(by Lemma 2.1.4),}
\end{aligned}$$

and the proposition is proved. □

We close this section with a further brief discussion of the relationship between the additive energy of a set and its doubling constant. Lemma 2.1.3 shows that sets of small doubling have to have large normalised additive energy. The converse to this does not hold: if $A \subset \mathbb{Z}$ is the union of an arithmetic progression of length n and a suitable random set of size n then A will have large normalised additive energy by Lemma 2.1.3, and also large doubling by Proposition 2.1.1. Note, however, that although this set A does not have small doubling, a large subset of it does have small doubling. It turns out that this is a general phenomenon, as is shown by the following result.

Theorem 2.1.5 (Balog–Szemerédi–Gowers) *Let $K \geq 1$. Let A be a finite subset of an abelian group G, and suppose that $\omega(A) \geq 1/K$. Then there exists a subset $A' \subset A$ satisfying $|A'| \gg K^{-O(1)}|A|$ and $|A'+A'| \ll K^{O(1)}|A'|$.*

When $G = \mathbb{Z}$, the qualitative statement of Theorem 2.1.5 is due to Balog and Szemerédi [2], and the bounds stated above were obtained by Gowers [30, Proposition 12]. Although Theorem 2.1.5 is an extremely useful result, we do not need it in this book, so we omit the full proof, instead noting only that it follows from [68, Theorem 2.31] and Theorem 2.3.1, below. We also note that Tao [62, Theorem 5.2] has obtained a version of Theorem 2.1.5 for an arbitrary group G with additive energy replaced by multiplicative energy.

2.2 Sets of Very Small Doubling

Classifying the sets of doubling at most K is difficult in general. However, for very small values of K it turns out that one can solve the problem completely using elementary combinatorial methods. To illustrate this, and to give a feel for the kinds of results we will look to prove more generally, in this section we prove the following theorem.

Theorem 2.2.1 (Freiman [27]) *Let G be a group and let $A \subset G$ be a finite subset such that $|A^2| < \frac{3}{2}|A|$. Then there exists a subgroup H with $|H| = |A^2|$ such that for every $a \in A$ we have $A \subset aH = Ha$.*

The key takeaway from this result is that A is a large proportion of a coset aH of a finite subgroup H, in the sense that $A \subset aH$ with $|A| \geq \frac{1}{K}|aH|$.

Remark Conversely, it is easy to see that if H is a finite subgroup of G and $x \in G$, and if $A \subset xH = Hx$ with $|A| \geq \frac{1}{K}|H|$, then $|A^2| \leq K|A|$. Thus Theorem 2.2.1 gives a complete classification of the sets of doubling less than $\frac{3}{2}$.

The proof we give of Theorem 2.2.1 is from Tao's blog [64]. We begin with the following lemma, where we identify the subgroup H and prove a preliminary bound on its size.

Lemma 2.2.2 *Let G be a group and let $A \subset G$ be a finite subset such that*

$$|A^2| < \tfrac{3}{2}|A|. \tag{2.2.1}$$

Then $H = A^{-1}A$ is a subgroup of G. Moreover, $H = AA^{-1}$ and $|H| < 2|A|$.

Proof Given $a, b \in A$, the bound (2.2.1) implies that $|aA \cap bA| > \frac{1}{2}|A|$, and hence that there exist more than $\frac{1}{2}|A|$ distinct pairs $(w, x) \in A \times A$ such that $aw = bx$, and hence $a^{-1}b = wx^{-1}$. This implies in particular that $A^{-1}A \subset AA^{-1}$, and also, replacing A by A^{-1}, that $AA^{-1} \subset A^{-1}A$ and hence $A^{-1}A = AA^{-1}$, as required. Moreover, since there are precisely $|A|^2$ pairs $(w, x) \in A \times A$, it also implies that

$$|A^{-1}A| < \frac{|A|^2}{\frac{1}{2}|A|} \leq 2|A|,$$

as required.

Since $A^{-1}A$ is automatically symmetric, to show that it is a subgroup it remains to prove that it is closed under the group operation. Given

$c, d \in A$ in addition to a and b, there are similarly more than $\frac{1}{2}|A|$ distinct pairs $(y, z) \in A \times A$ such that $c^{-1}d = yz^{-1}$, and so for at least one of the pairs (w, x) and one of the pairs (y, z) we must have $y = x$. This implies that $a^{-1}bc^{-1}d = wz^{-1} \subset AA^{-1} = A^{-1}A$; since a, b, c, d are all arbitrary elements of A, it follows that $A^{-1}A$ is indeed closed under the group operation, as required. □

In the next lemma we improve the bound on the size of H to that required by Theorem 2.2.1.

Lemma 2.2.3 *Let G be a group and let $A \subset G$ be a finite subset such that $|A^2| < \frac{3}{2}|A|$. Set $H = A^{-1}A$ and let $a \in A$. Then $A^2 = aHa$. In particular, $|H| = |A^2|$.*

Proof We trivially have

$$A \subset aH \cap Ha, \tag{2.2.2}$$

and so certainly $A^2 \subset aHa$. To prove the reverse inclusion, let $z \in aHa$ be arbitrary. Since H is a subgroup of G by Lemma 2.2.2, z has $|H|$ representations of the form xy with $x \in aH$ and $y \in Ha$. More than half of these x are in A by Lemma 2.2.2 and (2.2.2), as are more than half of these y, and so there must be at least one pair $x \in aH$ and $y \in Ha$ such that $z = xy$ and such that *both* $x, y \in A$, giving $z = xy \in A^2$. It follows that $aHa \subset A^2$, and hence $A^2 = aHa$, as claimed. □

Proof of Theorem 2.2.1 The set $H = A^{-1}A$ is a subgroup of G by Lemma 2.2.2. The fact that $A \subset aH$ for every $a \in A$ is trivial by definition of H, and the fact that $|H| = |A^2|$ is given by Lemma 2.2.3. It therefore remains to show that $aHa^{-1} = H$ for every $a \in A$.

Lemma 2.2.2 implies that $H = AA^{-1}$, and so for any given $a \in A$ we have $Aa^{-1} \subset aHa^{-1} \cap H$, and hence $|aHa^{-1} \cap H| \geq |A| > \frac{1}{2}|H|$ by Lemma 2.2.2. Since the only subgroup of H of cardinality greater than $\frac{1}{2}|H|$ is H itself, it follows that $aHa^{-1} = H$, which is to say that $aH = Ha$, as required. □

2.3 Iterated Sum Sets and the Plünnecke–Ruzsa Inequalities

In this section we present one of the fundamental tools in the study of sets of small doubling in abelian groups, the so-called *Plünnecke–Ruzsa inequalities*. These inequalities essentially say that if an abelian set has small doubling then it also has small tripling, small quadrupling, and so on, as follows.

Theorem 2.3.1 (Plünnecke–Ruzsa inequalities) *Let G be an abelian group, and let A, B be finite subsets of G. Suppose that $|A+B| \leq K|A|$. Then $|mB - nB| \leq K^{m+n}|A|$ for all non-negative integers m, n. In particular, if $|A + A| \leq K|A|$ then $|mA - nA| \leq K^{m+n}|A|$ for all non-negative integers m, n.*

In the next section we will illustrate the power of this theorem with an immediate application. Before seeing this or even proving the theorem, however, let us give a brief heuristic discussion of why one might expect such a result to be useful.

Consider first what it means for a subgroup $H < G$ to be closed under products and inverses. Of course, it means that if $h_1, h_2 \in H$ then h_1h_2, h_1^{-1} and h_2^{-1} all belong to H. However, what is also vital is that this can then be iterated, so that given any sequence $h_1, \ldots, h_m \in H$ and any $\epsilon_1, \ldots, \epsilon_m \in \{\pm 1\}$ we have $h_1^{\epsilon_1} \cdots h_m^{\epsilon_m} \in H$.

Theorem 2.3.1 gives us a way to iterate similarly the 'approximate closure' encoded in the small-doubling property, and thus to retain at least some control over the expressions $a_1 + \cdots + a_m - a'_1 - \cdots - a'_n$ in elements $a_i, a'_j \in A$. Indeed, we have $a_1 + \cdots + a_m - a'_1 - \cdots - a'_n \in mA - nA$. Theorem 2.3.1 then shows on the one hand that $mA - nA$ is not too much bigger than A, and on the other that $mA - nA$ is itself a set of small doubling, in the sense that $|2(mA - nA)| \leq K^{2m+2n}|mA - nA|$. Thus, when studying a set of small doubling in an abelian group, one can perform any bounded sequence of group operations without leaving the small-doubling regime, all the time remaining within a set that is reasonably comparable to the set one started with. This is an important part of why the theory of sets of small doubling works so well.

The fact that $|A+A| \leq K|A|$ implies that $|mA| \leq K^m|A|$ for all $m \in \mathbb{N}$ was first proved by Plünnecke [47]. This was subsequently rediscovered and extended to Theorem 2.3.1 by Ruzsa [53, 54]. Part of the proof was then dramatically simplified by Petridis [45], using the following lemma.

Lemma 2.3.2 (Petridis [45, Proposition 2.1]) *Let A and B be finite subsets of a group G, and let $U \subset A$ be a non-empty subset of A that minimises the ratio $|UB|/|U|$. Then, writing $R = |UB|/|U|$, for every finite subset $C \subset G$ we have*

$$|CUB| \leq R|CU|. \tag{2.3.1}$$

Proof We follow the exposition of Petridis's argument given in Gowers's blog post [32] (which Gowers attributes partly to Balog). When $C = \varnothing$ both sides of (2.3.1) are zero and the lemma holds. We may therefore assume that there exists some element $x \in C$. Writing $C' = C \setminus \{x\}$, we may assume by induction that

$$|C'UB| \leq R|C'U|. \tag{2.3.2}$$

Writing W for the set of all $u \in U$ such that $xu \in C'U$, we have

$$CU = C'U \cup (xU \setminus xW).$$

Since this is a disjoint union we may conclude that

$$|CU| = |C'U| + |U| - |W|. \tag{2.3.3}$$

Now $xW \subset C'U$ by definition of W, so $xWB \subset C'UB$. This implies that

$$CUB \subset C'UB \cup (xUB \setminus xWB),$$

and hence that

$$|CUB| \leq |C'UB| + |UB| - |WB|. \tag{2.3.4}$$

However, $|C'UB| \leq R|C'U|$ by (2.3.2), and $|UB| = R|U|$ by definition of R. Moreover, $|WB| \geq R|W|$ (by minimality of $|UB|/|U|$ if $W \neq \varnothing$, or trivially if $W = \varnothing$), so we may conclude from (2.3.4) that $|CUB| \leq R(|C'U| + |U| - |W|)$, and hence from (2.3.3) that $|CUB| \leq R|CU|$, as required. □

Armed with Lemma 2.3.2 we can easily deduce the following strengthening of a special case of Theorem 2.3.1.

Proposition 2.3.3 *Let A and B be finite subsets of an abelian group G, and let U and R be as in Lemma 2.3.2. Let $n \in \mathbb{N}$. Then $|U + nB| \leq R^n|U|$.*

Proposition 2.3.3 says in particular that if $|A + B| \leq K|A|$ as in Theorem 2.3.1 then, since $R \leq K$ by definition, we have $|nB| \leq K^n|A|$, which proves Theorem 2.3.1 in the case where we take only positive sums of the set B.

Proof The case $n = 1$ holds by definition of R, so we may assume that $n \geq 2$ and, by induction, that

$$|U + (n-1)B| \leq R^{n-1}|U|. \tag{2.3.5}$$

Lemma 2.3.2 then implies that $|U + nB| \leq R|U + (n-1)B|$, and so the desired inequality follows from (2.3.5). □

To pass from Proposition 2.3.3 to Theorem 2.3.1 we use a fundamental tool called the *Ruzsa triangle inequality*, one version of which runs as follows.

Lemma 2.3.4 (Ruzsa triangle inequality [52]) *Let U, V, W be subsets of a group. Then there exists an injection $\varphi : U \times V^{-1}W \to UV \times UW$. In particular, if U, V, W are finite then $|U||V^{-1}W| \leq |UV||UW|$.*

Proof We may define maps $v : V^{-1}W \to V$ and $w : V^{-1}W \to W$ in such a way for every $x \in V^{-1}W$ we have $x = v(x)^{-1}w(x)$, and then set $\varphi(u,x) = (uv(x), uw(x))$. To see that φ is injective, first note that $(uv(x))^{-1}(uw(x)) = v(x)^{-1}w(x) = x$, so that x is uniquely determined by $\varphi(u,x)$, and then note that $(uv(x))v(x)^{-1} = u$, so that u is uniquely determined by $\varphi(u,x)$ and x. □

Remark The term *triangle inequality* for Lemma 2.3.4 is motivated by the fact that the conclusion $|U||V^{-1}W| \leq |UV||UW|$ for all finite U, V, W is equivalent to the statement that

$$\log \frac{|V^{-1}W|}{|V|^{1/2}|W|^{1/2}} \leq \log \frac{|U^{-1}V|}{|U|^{1/2}|V|^{1/2}} + \log \frac{|U^{-1}W|}{|U|^{1/2}|W|^{1/2}}$$

(simply apply Lemma 2.3.4 with U^{-1} in place of U). The quantity $\log \frac{|V^{-1}W|}{|V|^{1/2}|W|^{1/2}}$ is often referred to in the literature as the *Ruzsa distance* $d(V,W)$ between V and W, and in some sense represents how 'incompatible' the multiplicative structures of V and W are. Lemma 2.3.4 therefore says that if U is 'somewhat compatible' with both V and W then V and W must be 'somewhat compatible' with one another. Note that this is not a true metric, since $d(V,W) = 0$ when V and W are two cosets of the same subgroup, and $d(V,V) \neq 0$ when V is not a coset of a subgroup.

Proof of Theorem 2.3.1 Let U and R be as in Lemma 2.3.2. It follows from Lemma 2.3.4 that $|U||mB - nB| \leq |U + mB||U + nB|$, and hence from Proposition 2.3.3 that $|U||mB - nB| \leq R^{m+n}|U|^2$, or equivalently that $|mB - nB| \leq R^{m+n}|U|$. The theorem then follows from the fact that $R \leq K$ and $|U| \leq |A|$. □

2.4 Ruzsa's Covering Argument

This section serves a number of connected purposes. One is to follow up Theorem 2.2.1 with another special setting in which we can quickly classify all the sets of small doubling, and hence add to the reader's sense of the kind of result we are aiming to prove in this book and this subject.

Theorem 2.4.1 (Ruzsa) *Let p be a prime, let $n \in \mathbb{N}$, and let A be a finite subset of the vector space $\mathbb{F}_p^n$. Suppose that $|A+A| \leq K|A|$. Then there exists a subspace H of $\mathbb{F}_p^n$ of cardinality at most $p^{K^4}K^2|A|$ such that $A \subset H$.*

Remark Theorem 2.4.1 is not entirely satisfactory, in that the bound $p^{K^4}K^2$ becomes increasingly bad as $p \to \infty$. In Theorem 4.1.2 we will obtain, with considerably more effort and at the expense of placing A inside more general sets than subspaces, a similar result in which the bounds are uniform over all abelian groups.

However, perhaps the main purpose of this section is to highlight the use of two fundamental tools in the proof of Theorem 2.4.1. The first of these tools is the Plünnecke–Ruzsa inequalities, which we stated and proved in the last section. We have already given some philosophical justification for the importance of these inequalities; Theorem 2.4.1 will now give a concrete illustration of it.

The second tool we wish to highlight, and which is the main focus of this section, is a technique called *covering*. This technique was introduced by Ruzsa, in fact in the proof of Theorem 2.4.1. As we shall see very shortly, covering allows us to prove the following slightly weakened version of Theorem 2.4.1.

Proposition 2.4.2 (Ruzsa [56]) *Let A be a finite subset of the vector space $\mathbb{F}_p^\infty$, and suppose that $|2A-2A| \leq K|A|$. Then there exists a subspace H of $\mathbb{F}_p^\infty$ of cardinality at most $p^K|A-A|$ such that $A \subset H$. In particular, $|H| \leq p^K K|A|$.*

It is then a straightforward matter to combine this with the Plünnecke–Ruzsa inequalities to prove Theorem 2.4.1, as follows.

Proof of Theorem 2.4.1 Theorem 2.3.1 implies that $|2A-2A| \leq K^4|A|$ and $|A-A| \leq K^2|A|$, and so the result follows from Proposition 2.4.2. □

It also turns out that the covering argument we use to prove Proposition 2.4.2 can be used to deduce the following corollary, which for large m is a significant improvement on the bound given by the Plünnecke–Ruzsa inequalities.

Corollary 2.4.3 (Green–Ruzsa [34]) *Let G be an abelian group, and let A be a finite subset of G. Suppose that $|A + A| \leq K|A|$. Then for every $m \in \mathbb{N}$ we have $|mA| \leq K^2 m^{K^4} |A|$.*

We therefore also take the opportunity to prove this corollary, giving a further illustration of the power of covering.

The most basic covering result, which underlies the general principle of covering, is the following powerful observation of Ruzsa.

Lemma 2.4.4 (Ruzsa's covering lemma) *Let A and B be finite subsets of some group. If $|AB|/|B| \leq K$ then there exists a subset $X \subset A$ with $|X| \leq K$ such that $A \subset XBB^{-1}$. Indeed, these properties are satisfied by taking X to be a subset of A that is maximal with respect to the property that the translates xB with $x \in X$ are all disjoint.*

Proof The disjointness of the translates xB implies that $|XB| = |X||B|$, and since $X \subset A$ we have $|XB| \leq |AB|$, and so $|X| \leq K$ as claimed. To prove that $A \subset XBB^{-1}$, note that if $a \in A$ then by the maximality of X there exists $x \in X$ such that aB and xB have non-empty intersection, which implies in particular that $a \in xBB^{-1}$. □

Remark 2.4.5 Essentially the same proof shows that if $|BA|/|B| \leq K$ then there exists a subset $Y \subset A$ with $|Y| \leq K$ such that $A \subset B^{-1}BY$. Alternatively, note that if $|BA|/|B| \leq K$ then $|A^{-1}B^{-1}|/|B^{-1}| \leq K$, and so Lemma 2.4.4 implies that there exists a set $X \subset A^{-1}$ of size at most K such that $A^{-1} \subset XB^{-1}B$, and hence $A \subset B^{-1}BX^{-1}$. We may therefore take $Y = X^{-1}$.

The term 'covering' comes from the fact that Lemma 2.4.4 says that A can be covered by a few translates of BB^{-1}. More generally, the term 'covering argument' tends to refer to any argument in which one starts with a hypothesis of the form $|A| \leq K|B|$ for some sets A, B, and strengthens this to a statement of the form 'A is covered by at most $O_K(1)$ translates of B'.

Covering enters the proof of Proposition 2.4.2 via the following lemma.

Lemma 2.4.6 (Ruzsa) *Let A be a finite subset of a group and suppose that $|A^{-1}A^2A^{-1}| \leq K|A|$. Then there exists $X \subset A^{-1}A^2$ with $|X| \leq K$ such that $A^{-1}A^n \subset X^{n-1}A^{-1}A$ for every $n \in \mathbb{N}$.*

Remark 2.4.7 In particular, if $|A^{-1}A^2A^{-1}| \leq K|A|$ then $|A^n| \leq K^n|A|$ for every $n \in \mathbb{N}$.

Proof It follows from Lemma 2.4.4 that there exists $X \subset A^{-1}A^2$ with $|X| \leq K$ such that

$$A^{-1}A^2 \subset XA^{-1}A. \tag{2.4.1}$$

For $n > 3$, we therefore have

$$\begin{aligned} A^{-1}A^n &= A^{-1}A^{n-1}A \\ &\subset X^{n-2}A^{-1}A^2 && \text{(by induction on } n\text{)} \\ &\subset X^{n-1}A^{-1}A && \text{(by (2.4.1))}, \end{aligned}$$

as claimed. □

Proof of Proposition 2.4.2 Lemma 2.4.6 implies that there exists a set $X \subset \mathbb{F}_p^\infty$ of size at most K such that for every n we have $nA - A \subset (n-1)X + A - A$; in particular, we have $\langle A \rangle \subset \langle X \rangle + A - A$. Since $|X| \leq K$ we have $|\langle X \rangle| \leq p^K$, and so we conclude that $|\langle A \rangle| \leq p^K|A - A|$. The theorem then follows from taking $H = \langle A \rangle$. □

Proof of Corollary 2.4.3 Theorem 2.3.1 implies that $|2A - 2A| \leq K^4|A|$, and so Lemma 2.4.6 implies that there is a set $X \subset G$ of size at most K^4 such that $mA - A \subset (m-1)X + A - A$, and hence $|mA| \leq |(m-1)X||A - A|$, which is in turn at most $K^2|(m-1)X||A|$ by Theorem 2.3.1. Writing $X = \{x_1, \ldots, x_r\}$, we have

$$(m-1)X \subset \{\ell_1 x_1 + \cdots + \ell_r x_r : 0 \leq \ell_i \leq m-1\},$$

and hence $|(m-1)X| \leq m^r \leq m^{K^4}$, and the result follows. □

Astute readers will have noticed that the conclusion of Lemma 2.4.6, and particularly the step (2.4.1) in its proof, are somewhat reminiscent of the definition of an approximate group. Lemma 2.4.6 can indeed be thought of as the first point in this book at which approximate groups, as opposed to small doubling, begin to play a role behind the scenes. Exercise 2.7 makes this role slightly more explicit.

Remark 2.4.8 In Green and Ruzsa's paper [34], which contains Corollary 2.4.3, the authors explicitly make the point that results such as

Lemma 2.4.6, Proposition 2.4.2 and Corollary 2.4.3 illustrate the utility of having a 'covering' property such as (2.4.1) rather than merely a bound on the size of a sum or product set. Since that paper and these results all pre-date the definition of an approximate group, it therefore seems fair to say that Green and Ruzsa anticipated that definition to some extent.

2.5 Small Tripling and Approximate Groups

The case $m = n = 1$ of Theorem 2.3.1 generalises without difficulty to non-abelian groups, as follows.

Lemma 2.5.1 *Let G be an arbitrary group, and suppose that $A \subset G$ is a finite subset satisfying $|A^2| \leq K|A|$. Then $|A^{-1}A| \leq K^2|A|$ and $|AA^{-1}| \leq K^2|A|$.*

Proof The fact that $|A^{-1}A| \leq K^2|A|$ follows from applying Ruzsa's triangle inequality, Lemma 2.3.4, with $U = V = W = A$. The fact that $|AA^{-1}| \leq K^2|A|$ then follows from the first part with A^{-1} in place of A. □

Unfortunately, however, the following example shows that Theorem 2.3.1 does not carry over to non-abelian groups in general.

Example 2.5.2 Let H be a finite group, and let $G = H * \langle x \rangle$, the free product of H and the infinite cyclic group with generator x. Set $A = H \cup \{x\}$. Then $|A^2| \leq 3|A|$, but A^3 contains HxH, which has size $|H|^2$, which is comparable to $|A|^2$.

It turns out that we can rectify this by slightly strengthening the assumption of small doubling. One possible strengthening is to assume $|A^3| \leq K|A|$ in place of $|A^2| \leq K|A|$, as follows.

Proposition 2.5.3 *Let A be a subset of a group, let $m \geq 3$ and let $\epsilon_1, \ldots, \epsilon_m \in \{\pm 1\}$. Suppose that $|A^3| \leq K|A|$. Then $|A^{\epsilon_1} \cdots A^{\epsilon_m}| \leq K^{3(m-2)}|A|$. If A is symmetric then $|A^m| \leq K^{m-2}|A|$.*

Proof Replacing K by $K^{1/3}$ in the asymmetric statement, both statements become $|A^{\epsilon_1} \cdots A^{\epsilon_m}| \leq K^{m-2}|A|$.

Once we have the $m = 3$ case, the other cases are straightforward applications of the Ruzsa triangle inequality. Indeed, given $m \geq 4$ we have

$$|A||A^{\epsilon_1} \cdots A^{\epsilon_m}| \leq |AA^{-\epsilon_2}A^{-\epsilon_1}||AA^{\epsilon_3} \cdots A^{\epsilon_m}|$$

by Lemma 2.3.4, and hence, by the case $m = 3$ and induction,

$$\begin{aligned}|A^{\epsilon_1} \cdots A^{\epsilon_m}| &\leq |A|^{-1}K|A| \cdot K^{m-3}|A| \\ &= K^{m-2}|A|,\end{aligned}$$

as required.

The case $m = 3$ is true by assumption in the symmetric case, but requires a proof in the asymmetric case. Again, we make repeated use of the Ruzsa triangle inequality. First, we have $|A^{-3}| = |A^3| \leq K^{1/3}|A|$ by assumption. Applying Lemma 2.3.4 with $U = W = A$ and $V = A^2$, we obtain

$$|A||A^{-2}A| \leq |A^3||A^2| \leq K^{2/3}|A|^2,$$

and hence $|A^{-2}A| \leq K^{2/3}|A|$. Since $(A^{-2}A)^{-1} = A^{-1}A^2$, we also have $|A^{-1}A^2| \leq K^{2/3}|A|$, and then applying these last two bounds to A^{-1} instead of A gives $|A^2A^{-1}| \leq K^{2/3}|A|$ and $|AA^{-2}| \leq K^{2/3}|A|$. Applying Lemma 2.3.4 with $U = V = A$ and $W = AA^{-1}$ then gives

$$|A||A^{-1}AA^{-1}| \leq |A^2||A^2A^{-1}| \leq K|A|^2,$$

and hence $|A^{-1}AA^{-1}| \leq K|A|$. Finally, applying this last bound to A^{-1} instead of A gives $|AA^{-1}A| \leq K|A|$. This proves the $m = 3$ case of the asymmetric statement, completing the proof of the proposition. □

Proposition 2.5.3 motivates the following definition.

Definition 2.5.4 (small tripling) Given a finite subset A of a group we call the quantity $|A|^3/|A|$ the *tripling constant* of A. If the tripling constant of A is at most a given constant K then we often say simply that A is a *set of tripling at most* K, or even merely a *set of small tripling*.

Another possible way to strengthen the notion of small doubling, hinted at already by Remark 2.4.7 and the proof of Lemma 2.4.6, is to replace small doubling with the notion of being an approximate group. Indeed, as in the proof of Lemma 2.4.6, a straightforward induction shows that if $A^2 \subset XA$ then $A^n \subset X^{n-1}A$ for every $n \in \mathbb{N}$, and so if A is a K-approximate group then

$$|A^n| \leq K^{n-1}|A|. \tag{2.5.1}$$

The following result, a version of which was first proved by Tao [62, Corollary 3.11], shows that these two possible strengthenings are essentially equivalent.

Proposition 2.5.5 *Let A be a finite subset of an arbitrary group. If A is a K-approximate group then A has tripling at most K^2. Conversely, if A has tripling at most K then there exists an $O(K^9)$-approximate group B containing A and satisfying $|B| \leq 7K^2|A|$. Indeed, we may take $B = (A \cup \{1\} \cup A^{-1})^2$.*

Proof If A is a K-approximate group then its tripling constant is at most K^2 by (2.5.1). Conversely, suppose that A has tripling at most K. Abbreviating $\widehat{A} = A \cup \{1\} \cup A^{-1}$ and taking $B = \widehat{A}^2$, we have

$$B = \{1\} \cup A \cup A^{-1} \cup A^2 \cup A^{-2} \cup AA^{-1} \cup A^{-1}A. \tag{2.5.2}$$

Each of the sets in this union has size at most $K^2|A|$ by Lemma 2.5.1, so $|B| \leq 7K^2|A|$ claimed. To show that B is an approximate group, first note that

$$\widehat{A}^3 = \bigcup_{\epsilon_i \in \{-1,0,1\}} A^{\epsilon_1} A^{\epsilon_2} A^{\epsilon_3}.$$

Proposition 2.5.3 implies that $|A^{\epsilon_1} A^{\epsilon_2} A^{\epsilon_3}| \leq K^3|A|$ for all choices of $\epsilon_i \in \{-1, 0, 1\}$, and so it follows that

$$\begin{aligned} |\widehat{A}^3| &\leq O(K^3)|A| \\ &\leq O(K^3)|\widehat{A}|. \end{aligned}$$

It therefore follows from Exercise 2.7 that $(A \cup \{1\} \cup A^{-1})^2$ is an $O(K^9)$-approximate group, as claimed. □

It turns out that replacing the condition of small doubling with that of small tripling or being an approximate group essentially loses no generality even in the non-abelian setting, despite Example 2.5.2. Indeed, that example gives a clue as to why this is. Although the set A presented there does not have small tripling, a large subset of it, namely H, does have small tripling, and is even an approximate group; A is thus in some sense an approximate group up to a small error. In fact, this turns out to be a general property, thanks to the following result, which strengthens an earlier similar result proved by Tao when he introduced the notion of an approximate group in [62].

Theorem 2.5.6 *Let A be a finite subset of a group G and suppose that $|A^2| \leq K|A|$. Then there exists a subset $U \subset A$ with $|UA| \leq K|U|$ such that $|U| \geq |A|/K$ and $|U^m| \leq K^{m-1}|U|$ for every $m \in \mathbb{N}$. Moreover, writing $\widehat{U} = U \cup \{1\} \cup U^{-1}$, the set $\widehat{U}^2$ is an $O(K^{24})$-approximate group of size at most $7K^2|A|$, and there exist sets $X, Y \subset A$ of size at most K^2 such that $A \subset X\widehat{U}^2 \cap \widehat{U}^2 Y$.*

Thus, if A has small doubling then a large proportion of A is a set U of small tripling, and A can be covered by a few left translates or a few right translates of some $O(K^{O(1)})$-approximate group $\widehat{U}^2$ that is not too much larger than A.

Proof Let U be a non-empty subset of A that minimises the ratio $|UA|/|U|$, noting that

$$|UA| \leq K|U| \tag{2.5.3}$$

as required. Note also that $|UA| \geq |A|$, which combines with (2.5.3) to prove that

$$|U| \geq |A|/K, \tag{2.5.4}$$

as required. It follows from Lemma 2.3.2 that for every $m \in \mathbb{N}$ we have $|U^m A| \leq K|U^m|$, but since $U \subset A$ this implies that $|U^{m+1}| \leq K|U^m|$, from which it follows by induction that $|U^m| \leq K^{m-1}|U|$ for every $m \in \mathbb{N}$, as required.

Remark 2.4.5 and (2.5.3) combine to imply that there exists a set $Y \subset A$ of size at most K such that $A \subset U^{-1}UY$. Since

$$\begin{aligned} |AU| &\leq |A^2| && \text{(because } U \subset A) \\ &\leq K|A| && \text{(by assumption)} \\ &\leq K^2|U| && \text{(by (2.5.4))}, \end{aligned}$$

Lemma 2.4.4 also implies that there exists a set $X \subset A$ of size at most K^2 such that $A \subset XUU^{-1}$. In particular, $A \subset X\widehat{U}^2 \cap \widehat{U}^2 Y$, as required. The set U has tripling at most K^2, so Proposition 2.5.5 implies that $\widehat{U}^2$ is an $O(K^{24})$-approximate group, as required. Finally, abbreviating $\widehat{A} = A \cup \{1\} \cup A^{-1}$, we have

$$\begin{aligned} |\widehat{U}^2| &\leq |\widehat{A}^2| && \text{(since } U \subset A) \\ &\leq 7K^2|A| && \text{(by (2.5.2) and Lemma 2.5.1)}, \end{aligned}$$

as required. □

Remark The commutativity of G in Proposition 2.3.3 allowed us to bound iterated sum sets of B by induction, whereas in the proof of the first part of Theorem 2.5.6 above we were only able to bound powers of U.

We close this section by revisiting Example 2.5.2. The reason the set A there failed to have small tripling was because of the existence of an element $x \in A$ for which AxA was large. It turns out that this is the only possible obstruction to a set of small doubling having small tripling, thanks to the following result, which again is originally due to Tao but in which the explicit bounds are due to Petridis.

Theorem 2.5.7 ([45, Theorem 1.6]) *Let A be a finite subset of a group, and suppose that $|A^2| \leq K_1|A|$ and $|AxA| \leq K_2|A|$ for every $x \in A$. Then for every $m \geq 3$ we have $|A^m| \leq K_1^{8m-17}K_2^{m-2}|A|$.*

We do not use Theorem 2.5.7 in this book, so we direct the reader to Petridis's paper for its proof.

2.6 Stability of Approximate Groups under Basic Operations

Two familiar properties of exact subgroups is that they behave well under group homomorphisms and intersections, for example in the sense that the intersection of two subgroups is another subgroup. In this section we show that approximate groups and sets of small tripling also behave well under group homomorphisms and intersections.

The fact that a homomorphic image of an approximate subgroup is again an approximate subgroup is trivial: given $A, X \subset G$ and a homomorphism $\pi : G \to \Gamma$, say, if $A^2 \subset XA$ then $|\pi(X)| \leq |X|$ and $\pi(A)^2 \subset \pi(X)\pi(A)$. For sets of small tripling we have the following result.

Proposition 2.6.1 (stability of small tripling under group homomorphisms) *Let G, Γ be groups, let $\pi : G \to \Gamma$ be a homomorphism, let $A \subset G$ be a finite symmetric set, and let $m \in \mathbb{N}$. Then*

$$\frac{|\pi(A)^m|}{|\pi(A)|} \leq \frac{|A^{m+2}|}{|A|}.$$

In particular, if $|A^3| \leq K|A|$ then

$$|\pi(A)^3| \leq K^3|\pi(A)|$$

by Proposition 2.5.3.

We prove Proposition 2.6.1 following Helfgott [40, Lemma 7.4]. We start with the following observation.

Lemma 2.6.2 *Let G be a group with a subgroup H, and let $\pi : G \to G/H$ be the quotient map. Suppose that A is a finite symmetric subset of*

G. Then A is contained in the union of $|\pi(A)|$ left translates of $A^2 \cap H$. In particular,

$$|A^2 \cap H| \geq \frac{|A|}{|\pi(A)|}.$$

Note that there is no assumption that H is normal.

Proof If xH is a coset of H containing at least one element of A then we may assume without loss of generality that $x \in A$. If a is an arbitrary element of $A \cap xH$ then there exists $h \in H$ such that $a = xh$. It follows that $h = x^{-1}a \in A^2$, and hence $h \in A^2 \cap H$, and so $a \in x(A^2 \cap H)$ and the lemma is satisfied. □

The complement to Lemma 2.6.2 in the proof of Proposition 2.6.1 is the following lemma.

Lemma 2.6.3 *Let $m, n \in \mathbb{N}$. Let G be a group with a subgroup H, and let $\pi : G \to G/H$ be the quotient map. Suppose that A is a finite symmetric subset of G. Then $|\pi(A^m)||A^n \cap H| \leq |A^{m+n}|$.*

Proof Define a map $\varphi : \pi(A^m) \to A^m$ by choosing, for each $x \in \pi(A^m)$, an element $\varphi(x) \in A^m$ such that $\pi(\varphi(x)) = x$. We then have

$$\varphi(\pi(A^m))(A^n \cap H) \subset A^{m+n}$$

by definition of φ, and

$$\begin{aligned} |\varphi(\pi(A^m))(A^n \cap H)| &= |\varphi(\pi(A^m))||A^n \cap H| \\ &= |\pi(A^m)||A^n \cap H|, \end{aligned}$$

because $\varphi(\pi(A^m))$ contains at most one element in each left coset of H. The result follows. □

Proof of Proposition 2.6.1 Lemma 2.6.2 implies that

$$|A^2 \cap \ker \pi| \geq \frac{|A|}{|\pi(A)|},$$

whilst Lemma 2.6.3 implies that

$$|\pi(A^m)||A^2 \cap \ker \pi| \leq |A^{m+2}|.$$

The result follows. □

We now come onto results about intersections, starting with the following.

Proposition 2.6.4 (stability of small tripling under intersections with subgroups) *Let A be a finite symmetric subset of a group G containing the identity, and let H be a subgroup of G. Then*

$$\frac{|A^m \cap H|}{|A^2 \cap H|} \leq \frac{|A^{m+1}|}{|A|}$$

for every $m \in \mathbb{N}$. In particular, if $|A^3| \leq K|A|$ then

$$|(A^m \cap H)^3| \leq K^{3m-1}|A^m \cap H|$$

for every $m \geq 2$ by Proposition 2.5.3.

Proof Fix $m \in \mathbb{N}$ and write $\pi : G \to G/H$ for the quotient map. Then Lemma 2.6.2 then implies that $|A^2 \cap H| \geq |A|/|\pi(A)|$, whilst Lemma 2.6.3 implies that $|A^m \cap H| \leq |A^{m+1}|/|\pi(A)|$. □

In a similar direction, in Exercise 2.10 we invite the reader to show that if A and B are sets of small tripling in a group then $A^2 \cap B^2$ is also a set of small tripling.

Finally, we show that approximate subgroups behave well under intersections.

Proposition 2.6.5 (stability of approximate groups under intersections) *Let A be a K-approximate subgroup of a group G and B an L-approximate subgroup of G. Then for every $m, n \geq 2$ the set $A^m \cap B^n$ is covered by at most $K^{m-1}L^{n-1}$ left translates of $A^2 \cap B^2$. In particular, $A^m \cap B^n$ is a $K^{2m-1}L^{2n-1}$-approximate group, and if H is a subgroup of G then $A^m \cap H$ is a K^{2m-1}-approximate group.*

The proof of Proposition 2.6.5 rests on the following fact.

Lemma 2.6.6 *Let x, y be elements of a group G and let A, B be symmetric subsets of G such that $xA \cap yB \neq \varnothing$. Then there exists $z \in xA \cap yB$ such that $xA \cap yB \subset z(A^2 \cap B^2)$.*

Proof If there exists $z = xa = yb$ with $a \in A$ and $b \in B$ then $xA \cap yB \subset z(a^{-1}A \cap b^{-1}B) \subset z(A^2 \cap B^2)$. □

Proof of Proposition 2.6.5 The set A^m is covered by K^{m-1} left translates of A, whilst the set B^n covered by L^{n-1} left translates of B, and so $A^m \cap B^n$ is covered by at most $K^{m-1}L^{n-1}$ sets of the form $xA \cap yB$. The result therefore follows from Lemma 2.6.6. □

We invite the reader to show in Exercise 2.11 that Propositions 2.6.5 and 2.6.4 fail if $m = 1$ or $n = 1$.

2.7 Freiman Homomorphisms

Let G be a group, and let A be a finite subset of G. It is clear that in order to study the doubling of A it is sufficient to restrict attention to the subgroup $\langle A\rangle$ generated by A. For example, if $\langle A\rangle$ is abelian we can use abelian techniques to study $A + A$, even if the larger group G is not abelian.

It turns out that we can take this slightly further. For example, if $A = \{-n, \ldots, n\} \subset \mathbb{Z}$ then $\langle A\rangle = \mathbb{Z}$, but we do not really need to consider the whole of $\mathbb{Z}$ in order to understand the set $A + A$. Indeed, given a prime $p \geq 10n$, say, if we define instead $A = \{-n, \ldots, n\} \subset \mathbb{Z}/p\mathbb{Z}$ then the set $A + A$ looks exactly the same as it did when we defined A as a subset of $\mathbb{Z}$. In particular, its doubling constant is the same in each case. Thus, from the point of view of its doubling, it does not matter whether the set A is viewed as being a subset of $\mathbb{Z}$ or a subset of $\mathbb{Z}/p\mathbb{Z}$, provided p is large enough not to interfere with $A + A$.

Of course, the set $A = \{-n, \ldots, n\} \subset \mathbb{Z}/p\mathbb{Z}$ is a homomorphic image of $\{-n, \ldots, n\} \subset \mathbb{Z}$, but even this will not always give a means of passing between 'equivalent' sets in different groups. For example, if p' is some other large prime then there are no non-trivial group homomorphisms between $\mathbb{Z}/p\mathbb{Z}$ and $\mathbb{Z}/p'\mathbb{Z}$, but the set $A' = \{-n, \ldots, n\} \subset \mathbb{Z}/p'\mathbb{Z}$ is still equivalent to A in some sense.

The tool that allows us to pass between A and A' in this case is the *Freiman homomorphism*.

Definition 2.7.1 (Freiman homomorphism) Let A and B be subsets of groups, and let $k \in \mathbb{N}$. Then a map $\varphi : A \to B$ is a *Freiman homomorphism of order k*, or a *Freiman k-homomorphism*, if whenever $x_1, \ldots, x_k, y_1, \ldots, y_k \in A$ satisfy

$$x_1 \cdots x_k = y_1 \cdots y_k$$

we have

$$\varphi(x_1) \cdots \varphi(x_k) = \varphi(y_1) \cdots \varphi(y_k).$$

If $1 \in A$ and $\varphi(1) = 1$ then we say that φ is *centred*. If φ is a bijective Freiman k-homomorphism and its inverse is also a Freiman k-homomorphism, we say that φ is a *Freiman isomorphism of order k*, or a *Freiman k-isomorphism*. If φ is a Freiman k-isomorphism then the set $\varphi(A)$ is called a *Freiman k-model* of A.

Remark Every map is trivially a Freiman 1-homomorphism, so it is the cases $k \geq 2$ that are meaningful. More generally, the property of being a Freiman k-homomorphism gets stronger as k increases. Indeed, let $\varphi : A \to B$ be a Freiman k-homomorphism. Assuming without loss of generality that $A \neq \varnothing$, we may pick an arbitrary element $a \in A$, and then given $m < k$ and $x_1, \dots, x_m, y_1, \dots, y_m \in A$ with $x_1 \cdots x_m = y_1 \cdots y_m$ we also have $x_1 \cdots x_m a^{k-m} = y_1 \cdots y_m a^{k-m}$, from which it easily follows that $\varphi(x_1) \cdots \varphi(x_m) = \varphi(y_1) \cdots \varphi(y_m)$.

The equivalence of the sets A and A' defined above is captured formally by the fact that they are Freiman 2-isomorphic to one another. In Chapter 3 and particularly Chapter 4 we will make heavy use of the fact that subsets of infinite groups can in principle have Freiman-isomorphic images in finite groups.

We now present some basic properties of Freiman homomorphisms. We leave the proofs as exercises.

Lemma 2.7.2 (basic properties of Freiman homomorphisms) *Let A, B and C be subsets of groups, and let $k \in \mathbb{N}$. Then the following hold.*

(i) *If $\varphi : A \to B$ and $\psi : B \to C$ are Freiman k-homomorphisms then so is the composition $\psi \circ \varphi : A \to C$.*
(ii) *If $\varphi : A \to B$ is a centred Freiman 2-homomorphism and $a, a^{-1} \in A$ then $\varphi(a^{-1}) = \varphi(a)^{-1}$.*
(iii) *If H, G are groups and $\varphi : H \to G$ is a centred Freiman 2-homomorphism then φ is a group homomorphism.*
(iv) *Conversely, if $\varphi : A \to B$ is the restriction of a group homomorphism then φ is a centred Freiman k-homomorphism.*

Lemma 2.7.3 (basic properties of abelian Freiman homomorphisms) *Let A and B be subsets of abelian groups, and let $k \in \mathbb{N}$. Then the following hold.*

(i) *If $\varphi : A \to B$ is a Freiman k-homomorphism and $1 \leq \ell \leq k$ then for every $x_1, \dots, x_k, y_1, \dots, y_k \in A$ with*

$$x_1 + \cdots + x_\ell - x_{\ell+1} - \cdots - x_k = y_1 + \cdots + y_\ell - y_{\ell+1} - \cdots - y_k$$

we have

$$\varphi(x_1) + \cdots + \varphi(x_\ell) - \varphi(x_{\ell+1}) - \cdots - \varphi(x_k) = \varphi(y_1) + \cdots + \varphi(y_\ell) - \varphi(y_{\ell+1}) - \cdots - \varphi(y_k).$$

(ii) *If* $\varphi : A \to B$ *is a Freiman* k*-homomorphism and* $\ell + m \leq k$ *then the map* $\varphi_{\ell,m} : \ell A - mA \to \ell B - mB$ *given by*

$$a_1 + \cdots + a_\ell - a_1' - \cdots - a_m'$$
$$\mapsto \varphi(a_1) + \cdots + \varphi(a_\ell) - \varphi(a_1') - \cdots - \varphi(a_m').$$

is a well-defined Freiman $\lfloor k/(\ell + m) \rfloor$*-homomorphism.*

(iii) *If* $\varphi : A \to B$ *is a Freiman* k*-homomorphism then every translate of* φ *is also a Freiman* k*-homomorphism. In particular, if* $0 \in A$ *then* $\varphi - \varphi(0)$ *is a centred Freiman* k*-homomorphism. Similarly, if* $\varphi : A \to B$ *is a Freiman* k*-isomorphism then every translate of* φ *is also a Freiman* k*-isomorphism.*

Lemma 2.7.4 (stability of approximate groups under Freiman homomorphisms) *Let* A *be a* K*-approximate group, let* G *be a group, and let* $\varphi : A^3 \to G$ *be a centred Freiman* 2*-homomorphism. Then* $\varphi(A)$ *is a* K*-approximate group.*

Proof By definition there exists X with $|X| \leq K$ such that $A^2 \subset XA$. We may assume that X is a minimal set such that this holds, which implies in particular that for every $x \in X$ there exists $a, b, c \in A$ such that $ab = xc$, and hence that $X \subset A^3$. In particular, this implies that X is in the domain of φ, and so, given $a, b \in A$, there exist $x \in X$ and $c \in A$ such that $ab = xc$, and hence $\varphi(a)\varphi(b) = \varphi(x)\varphi(c)$. We therefore have $\varphi(A)^2 \subset \varphi(X)\varphi(A)$. The fact that φ is centred implies that $1 \in \varphi(A)$, and also, by Lemma 2.7.2 (ii), that $\varphi(A)$ is symmetric. This completes the proof. □

Exercises

2.1 Show that if A is a finite subset of a group such that $|A^2| < 2|A|$ then $A^{-1}A = AA^{-1}$.

2.2 Show that for a finite set A in an abelian group we have $|A| \leq |A + A| \leq \frac{|A|^2+|A|}{2}$, and for arbitrary $n \in \mathbb{N}$ give an example with $|A| = n$ where these bounds are attained. Deduce that

$$\frac{\liminf_{n\to\infty} \mathbb{E}[|A_{k,n} + A_{k,n}|]}{D(k)} \to 1,$$

as claimed in Remark 2.1.2.

2.3 Show that for a finite set $A \subset \mathbb{Z}$ we have $|A+A| \geq 2|A| - 1$, and show that this bound is attained if and only if A is an arithmetic progression.

2.4 Let A be a finite subset of an arbitrary group such that $|AA^{-1}A| < 2|A|$. Show that $H = AA^{-1}$ is a subgroup such that $A \subset Hx$ for every $x \in A$.

2.5 Does Lemma 2.5.1 have a converse? That is, does $|A^{-1}A| \leq K|A|$ or $|AA^{-1}| \leq K|A|$ imply a bound of the form $|A^2| \leq f(K)|A|$? For which values of $\epsilon_1, \epsilon_2, \epsilon_3 \in \{\pm 1\}$ does the bound $|A^{\epsilon_1}A^{\epsilon_2}A^{\epsilon_3}| \leq K|A|$ imply a bound of the form $|A^3| \leq f_{\epsilon_1,\epsilon_2,\epsilon_3}(K)|A|$?

2.6 Let A be a finite subset of an abelian group and suppose that $|A+A| \leq K|A|$. Show that $A - A$ is a K^5-approximate group.

2.7 Let A be a finite symmetric subset containing the identity in an arbitrary group. Show that if $|A^5| \leq K|A|$ then A^2 is a K-approximate group; that if $|A^4| \leq K|A|$ then A^2 is a K^2-approximate group; and that if $|A^3| \leq K|A|$ then A^2 is a K^3-approximate group.

2.8 Let A be a finite symmetric subset containing the identity in an arbitrary group. Show that if $|A^{2n+1}| \leq K|A^n|$ for some $n \in \mathbb{N}$ then there is an $O_K(1)$-approximate group B with $A^n \subset B$ and $|B| \leq K|A^n|$.

2.9 Let G be a group, let $A \subset G$ be a K-approximate group, and let $B \subset G$ be an L-approximate group. Let $m, n \in \mathbb{N}$ be even. Show that $A^m \cap B^n$ is a $K^{3m/2}L^{3n/2}$-approximate group. *This improves the last part of Lemma 2.6.5.*

2.10 Let A, B be finite symmetric subsets of a group G that contain the identity.

(a) Show that $\sum_{x \in A^{-1}B} |Ax \cap B| = |A||B|$.

(b) Show that $|A^m \cap B^n| \leq |A^{m+1}x \cap B^{n+1}|$ for every $m, n \in \mathbb{N}$ and $x \in A^{-1}B$.

(c) Show that

$$\frac{|A^m \cap B^n|}{|A^2 \cap B^2|} \leq \frac{|A^{m+1}|}{|A|} \frac{|B^{n+1}|}{|B|}$$

for every $m, n \geq 2$. *Hint: Use part (a) to show that* $|A^2 \cap B^2| \geq |A||B|/|AB|$, *and parts (a) and (b) to show that* $|AB||A^m \cap B^n| \leq |A^{m+1}||B^{n+1}|$.

(d) Deduce that if $|A^3| \leq K|A|$ and $|B^3| \leq L|B|$ then

$$|(A^m \cap B^n)^3| \leq K^{3m-1}L^{3n-1}|A^m \cap B^n|$$

for every $m, n \geq 2$.

2.11 Show that there exists $K \geq 1$ such that the following holds: if H is an arbitrary finite group and $B \subset H$ is an arbitrary subset then there exists a group G with $H < G$ and a subset $A \subset G$ with $|A^3| \leq K|A|$ such that $A \cap H = B$. Deduce that Proposition 2.6.4 and Exercise 2.10 do not necessarily hold if either of m or n is equal to 1. Adapt your example to show that Proposition 2.6.5 does not necessarily hold if either of m or n is equal to 1.

2.12 Let A be a K-approximate group, let G be a group, and let $\varphi : A \to G$ be a centred Freiman 3-homomorphism. Show that $\varphi(A)$ is a K-approximate group.

2.13 Prove Lemmas 2.7.2 and 2.7.3.

2.14 Let $k \in \mathbb{N}$ and let A be a finite subset of a torsion-free abelian group. Exhibit a Freiman k-isomorphism of A to a subset of a finite cyclic group.

3

Coset Progressions and Bohr Sets

3.1 Introduction

We noted in Section 1.1 that finite subgroups and small sets were trivial examples of sets of small doubling. In this chapter we present and develop some more interesting examples of sets of small doubling in abelian groups, starting with the following.

Definition 3.1.1 (coset progression) Given elements $x_1, \ldots, x_r$ of an abelian group G, and positive integers $L_1, \ldots, L_r$, we call the set

$$P(x_1, \ldots, x_r; L_1, \ldots, L_r) = \{\ell_1 x_1 + \cdots + \ell_r x_r : -L_i \leq \ell_i \leq L_i\}$$

a *generalised arithmetic progression*, or simply a *progression*, with *rank* r and *side lengths* $L_1, \ldots, L_r$. We define this progression to be *proper* if the elements $\ell_1 x_1 + \cdots + \ell_r x_r$ are distinct for distinct $(\ell_1, \ldots, \ell_r) \in [L_1]^{\pm} \times \cdots \times [L_r]^{\pm}$. We sometimes abbreviate $P(x_1, \ldots, x_d; L_1, \ldots, L_d)$ as $P(x; L)$ or $P(x; L_1, \ldots, L_d)$.

If, in addition, H is a finite subgroup of G we call the set $H + P(x; L)$ a *coset progression* of rank r. We define $H + P(x; L)$ to be *proper* if the elements $h + \ell_1 x_1 + \cdots + \ell_r x_r$ are distinct for distinct $(h, \ell_1, \ldots, \ell_r) \in H \times [L_1]^{\pm} \times \cdots \times [L_r]^{\pm}$.

Note that a finite subgroup of an abelian group is a coset progression of rank 0.

A useful way of thinking of progressions is as homomorphic images of 'boxes' in $\mathbb{Z}^r$. Indeed, given elements $x_1, \ldots, x_r$ of an abelian group G and $L_1, \ldots, L_r \in \mathbb{N}$, and writing $B = [-L_1, L_1] \times \cdots \times [-L_r, L_r] \subset \mathbb{R}^r$ and $\pi : \mathbb{Z}^r \to G$ for the unique homomorphism such that $\pi(e_i) = x_i$ for each i, we have $P(x; L) = \pi(\mathbb{Z}^r \cap B)$. This is illustrated in Figure 3.1.

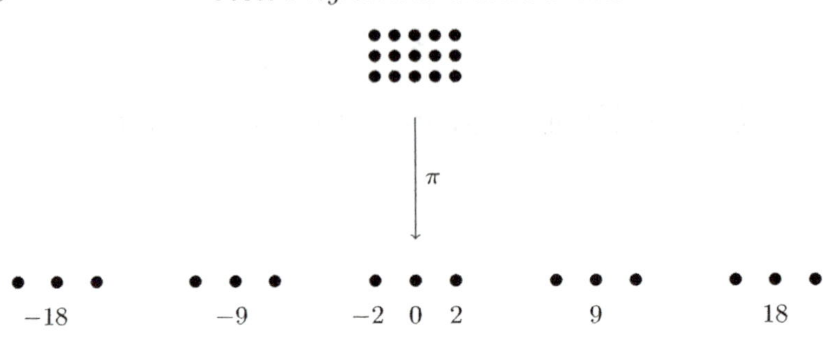

Figure 3.1 The progression $P(9,2;2,1) \subset \mathbb{Z}$ can be viewed as $\pi(\mathbb{Z}^2 \cap ([-2,2]\times[-1,1]))$, with $\pi : \mathbb{Z}^2 \to \mathbb{Z}$ defined via $\pi(1,0) = 9$ and $\pi(0,1) = 2$.

Lemma 3.1.2 (coset progressions are approximate groups) *Let $H + P = H + P(x;L)$ be a coset progression of rank r in an abelian group G. Let $k \in \mathbb{N}$. Then there exists a set $X \subset H + (k-1)P$ of size at most k^r such that $k(H+P) \subset X + H + P$. In particular, $H + P$ is a 2^r-approximate group, and $|k(H+P)| \le k^r|H+P|$ for every $k \in \mathbb{N}$.*

Proof Let $e_1,\ldots,e_r$ be the standard basis of $\mathbb{Z}^r$. Note that there exists a set $X_0 \subset (k-1)P(e;L)$ of size at most k^r such that $kP(e;L) \subset X + P(e;L)$; Figure 3.2 illustrates this in the case $r = 2$, $k = 4$. Writing $\pi : \mathbb{Z}^r \to G/H$ for the unique homomorphism such that $\pi(e_i) = H + x_i$, the lemma then follows by picking, for each $x \in X_0$, an element $x' \in \pi(x)$, and taking X to consist of these elements x'. □

It turns out that another way of producing sets of small doubling is via *inverse* images of boxes. To do this requires some notation. First, write $\mathbb{T} = \mathbb{R}/\mathbb{Z}$. Then, given $(x_1,\ldots,x_d) \in \mathbb{T}^d$, define $\|(x_1,\ldots,x_d)\|_{\mathbb{T}^d} \ge 0$ by writing $(\widehat{x}_1,\ldots,\widehat{x}_d)$ for the unique representative of $(x_1,\ldots,x_d)$ in $(-\frac{1}{2},\frac{1}{2}]^d$, and setting $\|(x_1,\ldots,x_d)\|_{\mathbb{T}^d} = \|(\widehat{x}_1,\ldots,\widehat{x}_d)\|_\infty$. Write $\widehat{G}$ for the space of homomorphisms $G \to \mathbb{T}$.

Definition 3.1.3 (Bohr set) Let G be a finite abelian group, let $d \in \mathbb{N}$, let $\gamma \in \widehat{G}^d$, and let $\rho \in [0,\frac{1}{2}]$. Then we call the set

$$B(\gamma,\rho) = \{x \in G : \|\gamma(x)\|_{\mathbb{T}^d} \le \rho\}$$

a *Bohr set* of rank d. In Chapter 4 it will be useful to use some alternative notation: given $\Gamma \subset \widehat{G}$ we write

$$B(\Gamma,\rho) = \{x \in G : \|\gamma(x)\|_{\mathbb{T}} \le \rho \text{ for every } \gamma \in \Gamma\}.$$

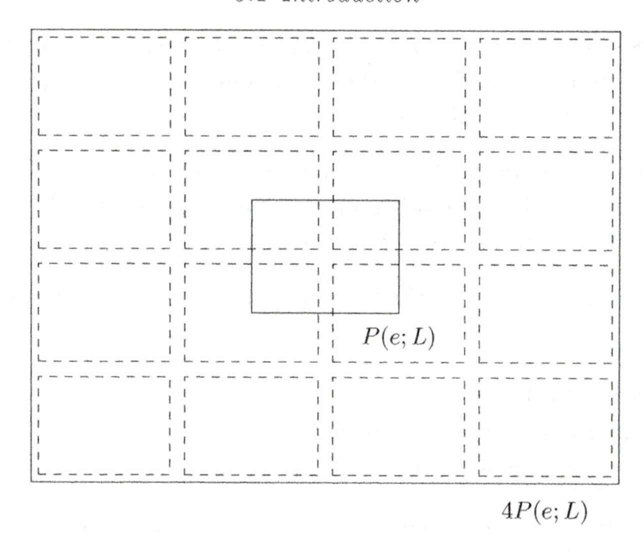

Figure 3.2 The sum set $4P(e; L)$ covered by 4^2 translates of $P(e, L)$ in the $r = 2$ case of Lemma 3.1.2.

Note that these two definitions give the same set if $\gamma = (\gamma_1, \ldots, \gamma_d)$ and $\Gamma = \{\gamma_1, \ldots, \gamma_d\}$.

Note that $B(\gamma, \rho)$ is the inverse image under γ of the cube $[-\rho, \rho]^d \subset \mathbb{T}^d$.

Proposition 3.1.4 (Bohr sets are approximate groups) *Let G be an abelian group, let $d \in \mathbb{N}$, let $\gamma \in \widehat{G}^d$, and let $\rho \leq \frac{1}{2}$. Then for every $k \in \mathbb{N}$ the set $kB(\gamma, \rho)$ is covered by $(2k)^d$ translates of $B(\gamma, \rho)$. In particular, $B(\gamma, \rho)$ is a 4^d-approximate group and $|kB(\gamma, \rho)| \leq (2k)^d |B(\gamma, \rho)|$.*

In the proof of Proposition 3.1.4 it will be helpful to define a slight variant of a Bohr set. Given $\gamma \in \widehat{G}^r$ and $\rho \leq \frac{1}{2}$, for every $\xi \in \mathbb{T}^r$ we define the *shifted Bohr set* $B(\gamma, \xi, \rho)$ via $B(\gamma, \xi, \rho) = \{x \in G : \|\gamma(x) - \xi\|_{\mathbb{T}^r} \leq \rho\}$. Thus $B(\gamma, 0, \rho) = B(\gamma, \rho)$, for example.

Lemma 3.1.5 *Let G be a finite abelian group, let $d \in \mathbb{N}$, let $\gamma \in \widehat{G}^d$, and let $\rho \leq \frac{1}{2}$. Let $\xi \in \mathbb{T}^d$. Then there exists $x_0 \in G$ such that*

$$B(\gamma, \xi, \tfrac{\rho}{2}) \subset B(\gamma, \rho) + x_0.$$

In particular, $|B(\gamma, \xi, \frac{\rho}{2})| \leq |B(\gamma, \rho)|$.

Proof See [68, Lemma 4.20]. If $B(\gamma, \xi, \frac{\rho}{2}) = \varnothing$ then the lemma is trivial with $x_0 = 0$. If $B(\gamma, \xi, \frac{\rho}{2}) \neq \varnothing$ then pick some $x_0 \in B(\gamma, \xi, \frac{\rho}{2})$ and note that $B(\gamma, \xi, \frac{\rho}{2}) - x_0 \subset B(\gamma, \rho)$. □

Proof of Proposition 3.1.4 Note that $kB(\gamma, \rho) \subset B(\gamma, k\rho)$. The proposition therefore follows from Lemma 3.1.5 and the fact that $B(\gamma, k\rho)$ can be covered by $(2k)^r$ sets of the form $B(\gamma, \xi, \frac{\rho}{2})$. □

We have so far identified the following sets of small doubling in abelian groups:

- sets of bounded size;
- coset progressions of bounded rank;
- Bohr sets of bounded rank.

Since these examples are all approximate groups, Lemma 2.7.4 shows that Freiman-homomorphic images of any of them are also approximate groups, and in particular sets of small doubling. Note also that, given a set of small doubling, we can always obtain further sets of small doubling simply by taking 'dense' subsets of the initial set. Indeed, given a finite set B and a subset $A \subset B$, we define the *density* of A in B to be $|A|/|B|$. Then, if $|B^2| \leq K|B|$ and $A \subset B$ with density $1/C$, it is easy to see that $|A^2| \leq CK|A|$. Thus dense subsets of any of the examples listed above are themselves sets of small doubling. We must therefore add to the above list:

- Freiman-homomorphic images of any of the above examples;
- dense subsets of any of the above examples.

The principal aim of this chapter is to prove various results expressing the different examples in this list in terms of one another. In the next section we show that both sets of bounded size and Freiman-homomorphic images of coset progressions are dense subsets of coset progressions. In the last two sections we prove the following theorem, which is the main result of this chapter.

Theorem 3.1.6 *Let G be an abelian group. Suppose that B_0 is a Bohr set of rank d in some finite abelian group and $\varphi : 3B_0 \to G$ is a centred Freiman 2-homomorphism, and write $B = \varphi(B_0)$. Then there exists a coset progression $H+P$ of rank at most $d+(8d)^{2d}$ such that $B \subset H+P \subset (2+(8d)^{2d})B$. In particular, by Lemmas 3.1.4 and 2.7.4, B has density at least $1/\exp(\exp(O(d^{O(1)})))$ in $H+P$.*

Thus, every example we have given so far of a set of small doubling in an abelian group can be realised as a dense subset of a coset progression. It turns out that this is a general phenomenon: in Chapter 4 we prove the *Freiman–Green–Ruzsa theorem*, Theorem 4.1.2, which states that every set of small doubling in an abelian group can be realised as a dense subset of a coset progression.

Remark At first sight it is somewhat unsatisfactory to have the double-exponential dependence on the rank d in the bound on the density in Theorem 3.1.6. However, we should really compare the density to the *doubling constant* of B, which by Lemma 3.1.4 is exponential in d. The rank of the coset progression given by Theorem 3.1.6 is thus of comparable order to the doubling constant, and its density in B is just a single exponential in the doubling constant. These bounds are comparable to the bounds that one would obtain using the more general Freiman–Green–Ruzsa theorem from the next chapter (Theorem 4.1.2).

Remark The reader is invited to show in Exercise 3.1 that a coset progression can always be realised as a Freiman image of a Bohr set. This can be seen as a strong converse to Theorem 3.1.6, and in conjunction with that theorem shows that Bohr sets and coset progressions are essentially equivalent notions.

3.2 Small Sets and Freiman Images of Coset Progressions

We can make an immediate reduction to the list of examples we gave at the end of the last section. Given a subset $A \subset G$ of size at most K, say $A = \{a_1, \ldots, a_r\}$ with $r \leq K$, we have

$$A \subset P(a_1, \ldots, a_r; 1, \ldots, 1). \qquad (3.2.1)$$

Thus A is contained with density at most 3^K in a progression of rank at most K. We may therefore remove sets of bounded size from the list.

Remark 3.2.1 It is important to note that whilst (3.2.1) allows us to reduce the list of examples of sets of small doubling from a *qualitative* perspective, from a *quantitative* perspective we have lost something, since the size and doubling constant of the progression given by (3.2.1) are both exponential in the doubling constant of the set we started with. If one is concerned with optimising bounds, it can therefore be useful to treat small sets as being separate from coset progressions; we give further details in Remark 4.1.8.

We can also ignore Freiman-homomorphic images of coset progressions, since they are themselves dense subsets of coset progressions, as follows.

Lemma 3.2.2 *Let G be an abelian group and let $H+P$ be a coset progression of rank r in some other abelian group. Suppose that $\varphi : H+P \to G$ is a Freiman 2-homomorphism. Then $\varphi(H+P)-\varphi(0)$ is also a coset progression of rank r. In particular, $\varphi(H+P)+\{0,-\varphi(0),-2\varphi(0)\}$ is a coset progression of rank $r+1$ containing $\varphi(H+P)$ as a subset of density at least $\frac{1}{3}$.*

Proof Writing $P = P(x_1,\ldots,x_r;L_1,\ldots,L_r)$, set $y_i = \varphi(x_i)-\varphi(0)$ for each $i=1,\ldots,r$. We claim that

$$\varphi(H+P) = \varphi(H) + P(y_1,\ldots,y_r;L_1,\ldots,L_r). \tag{3.2.2}$$

Since Lemma 2.7.2 (iii) and Lemma 2.7.3 (iii) imply that $\varphi(H)-\varphi(0)$ is a subgroup of G, this is sufficient. In fact, we prove that

$$\varphi(h+\ell_1x_1+\cdots+\ell_rx_r) = \varphi(h)+\ell_1y_1+\cdots+\ell_ry_r \tag{3.2.3}$$

whenever $h\in H$ and $|\ell_i|\le L_i$.

It follows from Lemma 2.7.2 (ii) that $-y_i = \varphi(-x_i)-\varphi(0)$. We may therefore, on replacing x_i by $-x_i$ and y_i by $-y_i$ where necessary, assume that $\ell_i\ge 0$ for each i in (3.2.3). Moreover, (3.2.3) holds trivially when $\ell_i=0$ for every i, so we may assume that $\ell_1+\cdots+\ell_r>0$. This implies in particular that there exists some $\ell_i>0$, and so by induction on $\ell_1+\cdots+\ell_r$ we may assume that

$$\begin{aligned}\varphi(h+\ell_1x_1+\cdots+(\ell_i-1)x_i+\cdots+\ell_rx_r)&\\ =\varphi(h)+\ell_1y_1+\cdots+(\ell_i-1)y_i+\cdots+\ell_ry_r.&\end{aligned}$$

It follows that

$$\begin{aligned}&\varphi(h+\ell_1x_1+\cdots+\ell_rx_r)+\varphi(0)\\ &\quad=\varphi(h+\ell_1x_1+\cdots+(\ell_i-1)x_i+\cdots+\ell_rx_r)+\varphi(x_i)\\ &\quad=\varphi(h)+\ell_1y_1+\cdots+\ell_iy_i+\cdots+\ell_ry_r+\varphi(0),\end{aligned}$$

which implies (3.2.3) and therefore the lemma. □

3.3 Lattices

We now come onto Theorem 3.1.6. Before proving it, we will need to develop our understanding of the structure of Bohr sets. Given a finite abelian group G and $\gamma \in \widehat{G}^d$, the image $\gamma(G)$ is a discrete subgroup of $\mathbb{T}^d$. Given $\rho \in [0, \frac{1}{2}]$, the Bohr set $B(\gamma, \rho)$ is then the pullback under γ of $[-\rho, \rho]^d \cap \gamma(G)$.

To study such sets we use a field called the *geometry of numbers.* In this section and the next we present a brief summary of those aspects of the geometry of numbers that we need in order to prove Theorem 3.1.6. Our treatment is based on Cassels [21], to which the reader may turn for a far more detailed account of the field.

A significant part of the geometry of numbers is concerned with interactions between *lattices* and *symmetric convex bodies* in $\mathbb{R}^d$. In this section we define lattices and introduce some of their properties; in the next we deal with symmetric convex bodies.

Definition 3.3.1 (lattice) Let $d \in \mathbb{N}$, and let V be a d-dimensional real vector space. A *lattice* $\Lambda \subset V$ is a group generated by a basis for V. Equivalently, Λ is a lattice if there exists a basis $x_1, \dots, x_d$ for V such that

$$\Lambda = \{\xi_1 x_1 + \cdots + \xi_d x_d : \xi_i \in \mathbb{Z} \text{ for each } i\}.$$

We call $x_1, \dots, x_d$ a *basis* for Λ. If $\Gamma \subset \Lambda$ is another lattice then we say that Γ is a *sublattice* of Λ and write $\Gamma < \Lambda$.

It is easy to see that a lattice $\Lambda \subset \mathbb{R}^d$ is *discrete*, in the sense that given an arbitrary element $v \in \Lambda$ there exists an open neighbourhood A of v such that $\Lambda \cap A = \{v\}$. It is also useful to note the following converse.

Lemma 3.3.2 *Let $d \in \mathbb{N}$, and suppose that Λ is a discrete subgroup of a d-dimensional real vector space V such that* $\operatorname{span}_{\mathbb{R}}(\Lambda) = V$. *Then Λ is a lattice in V.*

Proof If $d = 1$, assume without loss of generality that $V = \mathbb{R}$, and note that discreteness implies that there is a minimal positive element $v \in \Lambda$. It follows that $\Lambda = \langle v \rangle$, which proves the lemma in the case $d = 1$.

If $d > 1$, let $v_0 \in \Lambda$ be arbitrary, and then note that discreteness implies that there is a minimal $\lambda > 0$ such that $\lambda v_0 \in \Lambda$. Set $v = \lambda v_0$, write $W = \operatorname{span}_{\mathbb{R}}(v)$, and note that

$$\Lambda \cap W = \langle v \rangle. \tag{3.3.1}$$

We claim that $\Lambda/(\Lambda \cap W)$ is a discrete subgroup of V/W. Indeed, if $x, v_1, v_2, \ldots \in \Lambda$ and $w_1, w_2, \ldots \in \Lambda \cap W$ are such that $v_n - w_n \to x$ as $n \to \infty$ then the discreteness of Λ implies that $v_n = x + w_n$ for all large enough n, and hence that v_n is eventually constant modulo W. This implies that $\Lambda/(\Lambda \cap W)$ is discrete in V/W. Since $\operatorname{span}_{\mathbb{R}}(\Lambda/(\Lambda \cap W)) = V/W$, by induction we may conclude that $\Lambda/(\Lambda \cap W)$ is a lattice in V/W, which is to say generated by a basis for V/W. Adding v to this basis gives a basis for V, which by (3.3.1) is also a generating set for Λ, which completes the proof. □

The relevance of lattices to Theorem 3.1.6 arises thanks to the following lemma.

Lemma 3.3.3 *Let G be a finite abelian group, let $\gamma \in \widehat{G}^d$, and set $\Lambda = \gamma(G) + \mathbb{Z}^d \subset \mathbb{R}^d$. Then Λ is a lattice in $\mathbb{R}^d$.*

Proof Set $B = [0, 1)^d$, and note that B is a complete set of coset representatives for $\mathbb{Z}^d$ in $\mathbb{R}^d$. In particular, $\Lambda \cap B$ is a complete set of coset representatives for $\mathbb{Z}^d$ in Λ. Since $\mathbb{Z}^d$ has finite index (at most $|G|$) in Λ, this implies that $|\Lambda \cap B| < \infty$, from which it easily follows that Λ is discrete. Since Λ contains $\mathbb{Z}^d$ we have $\operatorname{span}_{\mathbb{R}}(\Lambda) = \mathbb{R}^d$, and so it follows from Lemma 3.3.2 that Λ is a lattice, as required. □

Let Λ be a lattice in $\mathbb{R}^d$ with basis $x_1, \ldots, x_d$, and consider the parallelopiped

$$P = \{\eta_1 x_1 + \cdots + \eta_d x_d : \eta_i \in [0, 1) \text{ for each } i\}.$$

We call P the *fundamental parallelopiped* for Λ with respect to the basis $x_1, \ldots, x_d$. Since $x_1, \ldots, x_d$ is by definition also a basis for $\mathbb{R}^d$, there exist unique functions $x : \mathbb{R}^d \to \Lambda$ and $p : \mathbb{R}^d \to P$ such that

$$v = x(v) + p(v) \tag{3.3.2}$$

for each $v \in \mathbb{R}^d$. In particular, $\mathbb{R}^d$ is the countable disjoint union of the sets $x + P$ with $x \in \Lambda$.

Write vol for Lebesgue measure normalised with respect to the standard basis of $\mathbb{R}^d$. Given elements $x_1, \ldots, x_d \in \mathbb{R}^d$, define the *determinant* $\det(x_1, \ldots, x_d)$ to be the determinant of the $d \times d$ matrix whose columns are the elements $x_1, \ldots, x_d$ expressed as column vectors with respect to the standard basis for $\mathbb{R}^d$. Note that if P is a fundamental parallelopiped for a lattice Λ with respect to a basis $x_1, \ldots, x_d$ then

$$\operatorname{vol}(P) = |\det(x_1, \ldots, x_d)|. \tag{3.3.3}$$

Proposition 3.3.4 *Let Λ be a lattice in $\mathbb{R}^d$ with basis $x_1, \ldots, x_d$, and Γ a sublattice with basis $y_1, \ldots, y_d$. Then*

$$\frac{|\det(y_1, \ldots, y_d)|}{|\det(x_1, \ldots, x_d)|} = [\Lambda : \Gamma].$$

Note that Proposition 3.3.4 implies in particular that if $x_1, \ldots, x_d$ and $y_1, \ldots, y_d$ are two bases of the same lattice Λ then $|\det(x_1, \ldots, x_d)| = |\det(y_1, \ldots, y_d)|$. Once we have proved the proposition , we may therefore define the *determinant* $\det(\Lambda)$ of a lattice Λ by setting $\det(\Lambda) = |\det(x_1, \ldots, x_d)|$ for an arbitrary basis $x_1, \ldots, x_d$ of Λ.

Proof Let P be a fundamental parallelopiped for Λ with respect to $x_1, \ldots, x_d$, and Q a fundamental parallelopiped for Γ with respect to $y_1, \ldots, y_d$. Let

$$\begin{array}{ll} x : \mathbb{R}^d \to \Lambda & \qquad p : \mathbb{R}^d \to P \\ y : \mathbb{R}^d \to \Gamma & \qquad q : \mathbb{R}^d \to Q \end{array}$$

be the unique functions satisfying

$$v = x(v) + p(v) = y(v) + q(v) \tag{3.3.4}$$

for each $v \in \mathbb{R}^d$ as in (3.3.2). We claim that

$$|p^{-1}(u) \cap Q| = [\Lambda : \Gamma] \tag{3.3.5}$$

for every $u \in P$. Since p is a translation on each set $x + P$ with $x \in \Lambda$, it is measure preserving on restriction to each such set, and so (3.3.5) will then imply that $p(Q) = P$ and $\mathrm{vol}(Q) = [\Lambda : \Gamma]\,\mathrm{vol}(P)$, which by (3.3.3) gives the desired result.

To prove (3.3.5), first note that the uniqueness of $y(v)$ and $q(v)$ in (3.3.4) implies that Q is a complete set of coset representatives for Γ in $\mathbb{R}^d$. It follows that for every $u \in \mathbb{R}^d$ the set $Q - u$ is also a complete set of coset representatives for Γ in $\mathbb{R}^d$. This implies that for every $u \in \mathbb{R}^d$ the set $\Lambda \cap (Q - u)$ is a complete set of coset representatives for Γ in Λ. This in turn implies in particular that $|\Lambda \cap (Q - u)| = [\Lambda : \Gamma]$, and hence that $|(\Lambda + u) \cap Q| = [\Lambda : \Gamma]$. However, if $u \in P$ then $(\Lambda + u) \cap Q$ is precisely $|p^{-1}(u) \cap Q|$, and so this gives (3.3.5), as claimed. □

A similar argument to the proof of Proposition 3.3.4 gives the following.

Lemma 3.3.5 (Blichfeldt [7]) *Let $d \in \mathbb{N}$. Let Λ be a lattice in $\mathbb{R}^d$, and let $A \subset \mathbb{R}^d$ be a measurable set. Suppose that*

$$(A - A) \cap \Lambda = \{0\}. \tag{3.3.6}$$

Then $\mathrm{vol}(A) \leq \det(\Lambda)$.

Proof Let $x_1, \ldots, x_d$ be a basis for Λ, write P for the corresponding fundamental parallelopiped, and define maps $x : \mathbb{R}^d \to \Lambda$ and $p : \mathbb{R}^d \to P$ as in (3.3.2). On restriction to each set of the form $x + P$ with $x \in \Lambda$ the map p is a translation, and hence measure preserving. Moreover, (3.3.6) implies that for every $u \in P$ we have $|p^{-1}(u) \cap A| \leq 1$. It follows that $\mathrm{vol}(A) \leq \mathrm{vol}(p(A)) \leq \mathrm{vol}(P) = |\det(x_1, \ldots, x_d)|$, as required. □

3.4 Convex Bodies

In studying the Bohr set $B(\gamma, \rho)$ we essentially study the interaction of the lattice coming from Lemma 3.3.3 with the cube $[-\rho, \rho]$. The only property of $[-\rho, \rho]$ that we will really need is that its interior, $(-\rho, \rho)$, is a so-called *symmetric convex body*. We now define this term, starting with the adjective *convex*.

Definition 3.4.1 (convex set) Let V be a finite-dimensional real vector space. A set $A \subset V$ is said to be *convex* if whenever $x, y \in A$ and $\rho \in (0, 1)$ the point $\rho x + (1 - \rho)y \in A$ as well.

Definition 3.4.2 (convex body) A *convex body* $B \subset \mathbb{R}^d$ is a non-empty bounded open convex set; it is *symmetric* if for every $x \in B$ we also have $-x \in B$.

The purpose of this short section is to record some elementary properties of convex bodies. We start by introducing and clarifying some notation. Throughout the rest of this chapter, whenever $A \subset \mathbb{R}^d$ and $\lambda \in \mathbb{R}$, we write λA for the *dilate*

$$\lambda A = \{\lambda a : a \in A\}.$$

Note that if $\lambda \in \mathbb{Z}$ then there is the potential for ambiguity here, since in general we have defined λA to be the iterated sum set $A + \cdots + A$.

When there is the danger of this, we write instead

$$\lambda \cdot A = \{\lambda a : a \in A\}$$

to distinguish the dilate from the iterated sum set.

The following lemma shows that in the setting of symmetric convex bodies this potential ambiguity is harmless.

Lemma 3.4.3 *Let $d \in \mathbb{N}$ and let $B \subset \mathbb{R}^d$ be a symmetric convex body. Then for every $k, \ell \in \mathbb{N}$ we have*

$$k \cdot B - \ell \cdot B = kB - \ell B = (k + \ell)B.$$

The proof of Lemma 3.4.3 is a straightforward exercise, and so we omit it. The same goes for the following lemma.

Lemma 3.4.4 *Let $d, k \in \mathbb{N}$. Suppose $B \subset \mathbb{R}^d$ is a convex body and $\pi : \mathbb{R}^d \to \mathbb{R}^k$ is a linear map. Then $\pi(B)$ is also a convex body, and if B is symmetric then so is $\pi(B)$.*

Lemma 3.4.5 *Let $d \in \mathbb{N}$. Suppose that $U, V < \mathbb{R}^d$ are complementary subspaces, in the sense that $\mathbb{R}^d = U \oplus V$, and write $\pi : \mathbb{R}^d \to V$ for the projection taking $(u, v) \in U \oplus V = \mathbb{R}^d$ to $u \in U$. Suppose $B \subset \mathbb{R}^d$ is a convex body. Then there is a continuous map $f : \pi(B) \to B$ that is a right inverse to π in the sense that $\pi \circ f$ is the identity on $\pi(B)$.*

Proof Set $k = \dim V$. We first prove the lemma in the case where $k = 1$, say $V = \operatorname{span}_{\mathbb{R}}(v_0)$ for some $v_0 \in \mathbb{R}^d$. Define functions $\varphi^+, \varphi^- : \pi(B) \to \mathbb{R}$ by setting

$$\varphi^+(u) = \sup\{\lambda \in \mathbb{R} : u + \lambda v_0 \in B\},$$
$$\varphi^-(u) = \inf\{\lambda \in \mathbb{R} : u + \lambda v_0 \in B\}.$$

We claim first that φ^+ and φ^- are continuous. Let $u \in \pi(B)$, noting that by definition there exists $\xi \in \mathbb{R}$ such that $u + \xi v_0 \in B$. The openness of B therefore implies that there exists an open neighbourhood N of u in U such that

$$u + N + \xi v_0 \subset B. \tag{3.4.1}$$

It then follows from (3.4.1), convexity and the definition of φ^+ that for every $\varepsilon \in (0, 1)$ and every $x \in \mathbb{N}$ we have

$$\varphi^+(u + \varepsilon x) \in \varphi^+(u) + [-\varepsilon(\varphi^+(u) - \xi), \varepsilon(\varphi^+(u) - \xi)],$$

and so φ^+ is continuous at u, as claimed. The proof that φ^- is continuous is essentially identical (alternatively, replacing v_0 by $-v_0$ puts $-\varphi^-$ in the role of φ^+).

Given $u \in \pi(B)$, it follows from convexity and the definitions of φ^+ and φ^- that

$$v + \tfrac{1}{2}(\varphi^+(u) + \varphi^-(u))v_0 \in B.$$

This implies we may define a function $f : \pi(B) \to B$ by $f(u) = \frac{1}{2}(\varphi^+(u) + \varphi^-(u))v_0$. This is trivially a right inverse to π, and is continuous by the continuity of φ^+ and φ^-, and so satisfies the requirements of the lemma in the case $k = 1$.

If $k > 1$, let $v_1, \dots, v_k$ be a basis for V. For each $j = 1, \dots, k$ set $V_j = \operatorname{span}_{\mathbb{R}}(v_j)$ and $U_j = \operatorname{span}_{\mathbb{R}}(v_{j+1}, \dots, v_k) + U$, and define $\pi_j : V_j \oplus U_j \to U_j$ by setting

$$\pi_j(\lambda_j v_j + \cdots + \lambda_k v_k + u) = \lambda_{j+1} v_{j+1} + \cdots + \lambda_k v_k + u$$

for every $\lambda_i \in \mathbb{R}$ and $u \in U$. By repeated application of Lemma 3.4.4, for each j the set $\pi_j \circ \cdots \circ \pi_1(B)$ is a convex body in U_j. By the case $k = 1$ of the lemma, for each j we may therefore define a continuous function

$$f_j : \pi_j \circ \cdots \circ \pi_1(B) \to \pi_{j-1} \circ \cdots \circ \pi_1(B)$$

that is a right inverse to π_j. Since

$$\pi = \pi_k \circ \cdots \circ \pi_1,$$

the function

$$f_1 \circ \cdots \circ f_k : \pi(B) \to B$$

is therefore a continuous right inverse to π, and so satisfies the requirements of the lemma. □

3.5 Successive Minima and Minkowski's Second Theorem

Given a symmetric convex body $B \subset \mathbb{R}^d$, we define the *successive minima* $\lambda_1, \dots, \lambda_d$ of B with respect to Λ via

$$\lambda_i = \inf\{\lambda > 0 : \dim \operatorname{span}_{\mathbb{R}}(\lambda B \cap \Lambda) \geq i\}.$$

Writing $\overline{B}$ for the closure of B, we may choose inductively a list

$$v_1, \dots, v_d \in \mathbb{Z}^d$$

of linearly independent vectors such that $v_1, \ldots, v_i \in \lambda_i \overline{B}$ for every i. We call such a list a *directional basis* for Λ with respect to B; note that for a given B and Λ, a directional basis may not be uniquely defined. We also caution that a directional basis for Λ with respect to B need not be a basis for Λ in the sense of Definition 3.3.1 (see Exercise 3.5).

The key result we will need from the geometry of numbers is the following.

Theorem 3.5.1 (Minkowski's second theorem) *Let $B \subset \mathbb{R}^d$ be a symmetric convex body and let Λ be a lattice. Write $\lambda_1 \leq \cdots \leq \lambda_d$ for the successive minima of B with respect to Λ. Then $\lambda_1 \ldots \lambda_d \operatorname{vol}(B) \leq 2^d \det(\Lambda)$.*

Minkowski's second theorem actually also includes the lower bound $\lambda_1 \cdots \lambda_d \operatorname{vol}(B) \geq \frac{2^n}{n!} \det(\Lambda)$, but in this book we need only the upper bound stated in Theorem 3.5.1. The reader interested in the lower bound can consult [21, §VIII.4.3] for a proof.

We prove Theorem 3.5.1 following Tao and Vu [68, §3.5], starting with a result they call the *squeezing lemma.*

Lemma 3.5.2 (squeezing lemma [68, Lemma 3.31]) *Let $d, k \in \mathbb{N}$, let $\mu \in (0, 1]$, let $B \subset \mathbb{R}^d$ be a convex body, and let $V < \mathbb{R}^d$ be a k-dimensional subspace. Suppose that $A \subset B$ is an open set. Then there exists an open subset $A' \subset B$ satisfying*

$$\operatorname{vol}(A') = \mu^k \operatorname{vol}(A) \tag{3.5.1}$$

and

$$(A' - A') \cap V \subset (\mu(A - A)) \cap V. \tag{3.5.2}$$

Proof Let U be a complementary subspace to V in $\mathbb{R}^d$, so that $\mathbb{R}^d = U \oplus V$. Define the projection $\pi : \mathbb{R}^d \to U$ by setting $\pi(u + v) = u$ for every $u \in U$ and $v \in V$.

Let $f : \pi(B) \to B$ be the continuous right inverse to π given by Lemma 3.4.5, and note that by the convexity of B we may set

$$\begin{array}{rcrcl} \Phi & : & B & \to & B \\ & & x & \mapsto & \mu x + (1 - \mu) f(\pi(x)). \end{array}$$

We claim that Φ is a homeomorphism from B to $\Phi(B)$. It is certainly continuous by the continuity of f and π; we will show that it has a continuous inverse $\Phi(B) \to B$. First note that, by definition of f, there exists a continuous map $\varphi : \pi(B) \to V$ such that $f(u) = u + \varphi(u)$ for every $u \in \pi(B)$. For every $u \in U$ and $v \in V$ with $u + v \in B$, it follows that

$$\Phi(u+v) = u + \varphi(u) + \mu(v - \varphi(u)). \tag{3.5.3}$$

It follows that the map $\Phi(B) \to B$ defined by $u+v \mapsto u+\varphi(u)+\mu^{-1}(v-\varphi(u))$ is an inverse to Φ. The continuity of φ ensures that this inverse is continuous, and so Φ is a homeomorphism, as required.

The set $A' = \Phi(A)$ is therefore open. It is also a subset of B by definition of Φ. The expression (3.5.3) for Φ shows that for each $u \in U$ the map Φ contracts the set $A \cap (u+V)$ by a factor of μ in every direction of V, so Fubini's theorem (Theorem 1.5.4) gives (3.5.1).

Finally, suppose that $y \in (A' - A') \cap V$. By definition there exist $x_1, x_2 \in A$ such that $y = \Phi(x_1) - \Phi(x_2)$. Writing each x_i in the form $x_i = u_i + v_i$ for some $u_i \in U$ and $v_i \in V$, we may conclude from (3.5.3) that

$$y = u_1 - u_2 + \varphi(u_1) - \varphi(u_2) + \mu(v_1 - v_2 - \varphi(u_1) + \varphi(u_2)). \tag{3.5.4}$$

However, the assumption that $y \in V$ then forces $u_1 = u_2$, which combined with (3.5.4) implies that

$$\begin{aligned} y &= \mu(v_1 - v_2) \\ &= \mu(x_1 - x_2) \\ &\in \mu(A - A), \end{aligned}$$

giving (3.5.2), as required. □

Proof of Theorem 3.5.1 We follow Tao and Vu [68, §3.5]. Fix a directional basis $v_1, \ldots, v_d$ for Λ with respect to B, and for each $i = 0, 1, \ldots, d$ set $V_i = \operatorname{span}_{\mathbb{R}}(v_1, \ldots, v_i)$ and $\Lambda_i = \Lambda \cap (V_i \setminus V_{i-1})$, noting that

$$\lambda_j B \cap \Lambda_j = \{0\} \tag{3.5.5}$$

by definition of λ_j and Λ_j.

Set $B_0 = \frac{\lambda_d}{2} B$. Starting with $A_0 = B_0$, apply Lemma 3.5.2 iteratively to obtain a sequence $A_0, A_1, \ldots, A_{d-1}$ of open subsets of the convex body B_0 such that

$$\operatorname{vol}(A_i) = \left(\frac{\lambda_i}{\lambda_{i+1}}\right)^j \operatorname{vol}(A_{i-1}) \tag{3.5.6}$$

and

$$(A_i - A_i) \cap V_i \subset \left(\frac{\lambda_i}{\lambda_{i+1}}(A_{i-1} - A_{j-1})\right) \cap V \tag{3.5.7}$$

for each i.

It is immediate from (3.5.6) and the definition of A_0 that

$$\mathrm{vol}(A_{d-1}) = \frac{\lambda_1 \dots \lambda_d \,\mathrm{vol}(B)}{2^d}. \tag{3.5.8}$$

For each j we have

$$\begin{aligned}
(A_{d-1} - A_{d-1}) \cap V_j &\subset \left(\tfrac{\lambda_j}{\lambda_d}(A_{j-1} - A_{j-1})\right) \cap V_j && \text{(by (3.5.7))}\\
&\subset \left(\tfrac{\lambda_j}{\lambda_d}(B_0 - B_0)\right) \cap V_j && \text{(since } A_{j-1} \subset B_0\text{)}\\
&\subset \left(\tfrac{\lambda_j}{\lambda_d}\left(\tfrac{\lambda_d}{2}B - \tfrac{\lambda_d}{2}B\right)\right) \cap V_j && \text{(by definition of } B_0\text{)}\\
&\subset (\lambda_j B) \cap V_j && \text{(by Lemma 3.4.3),}
\end{aligned}$$

which by (3.5.5) and the definition of Λ_j implies that $(A_{d-1} - A_{d-1}) \cap \Lambda_j = \{0\}$. Since this holds for all j, we conclude that

$$(A_{d-1} - A_{d-1}) \cap \Lambda = \{0\}.$$

Lemma 3.3.5 therefore combines with (3.5.8) to prove the theorem. □

3.6 Finding Dense Coset Progressions in Bohr Sets

In this section we prove Theorem 3.1.6. Slightly counterintuitively, the main step is to show that a Bohr set *contains* a progression as a dense subset, as follows.

Proposition 3.6.1 *Let G be a finite abelian group, let $d \in \mathbb{N}$, let $\gamma \in \widehat{G}^d$, and let $\rho \in (0, \frac{1}{6})$. Then $B(\gamma, \rho)$ contains a proper coset progression $H + P$ of rank at most d and size at least $|B(\gamma, \rho)|/(4d)^{2d}$.*

Once we have this, Ruzsa's covering argument allows us to obtain the desired containment in the opposite direction, as follows.

Proof of Theorem 3.1.6 It follows from Proposition 3.6.1 that B_0 contains a coset progression $H_0 + P_0$ of rank at most d and size at least $|B_0|/(4d)^{2d}$. Proposition 3.1.4 implies that $|B_0 + H_0 + P_0| \le 4^d |B_0|$, which means in particular that $|B_0 + H_0 + P_0| \le (8d)^{2d}|H_0 + P_0|$. Lemma 2.4.4 then implies that there exists a set $X \subset B_0$ of size at most $(8d)^{2d}$ such that $B_0 \subset X + (H_0 + P_0) - (H_0 + P_0)$, and hence

$$B \subset \varphi(X) + \varphi(H_0 + P_0) - \varphi(H_0 + P_0). \tag{3.6.1}$$

Since φ is centred, Lemma 3.2.2 implies that there exists a coset progression $H + P \subset G$ of rank at most d such that $\varphi(H_0 + P_0) = H + P$, and the inclusion (3.6.1) therefore becomes $B \subset \varphi(X) + H + 2P$. However, writing $\varphi(X) = \{x_1, \ldots, x_t\}$, and defining $P' = P(x_1, \ldots, x_t; 1, \ldots, 1)$ in a similar fashion to (3.2.1), we have $\varphi(X) \subset P'$, and hence

$$B \subset H + 2P + P' \subset \left(2 + (8d)^{2d}\right) B$$

by Lemma 2.7.2 (ii). Since $H + 2P + P'$ is a coset progression of rank at most $d + (8d)^{2d}$, the theorem is proved. □

Proposition 3.6.1 is immediate from the following two propositions.

Proposition 3.6.2 *Let G be a finite abelian group, let $d \in \mathbb{N}$, let $\gamma \in \widehat{G}^d$, and let $\rho \in (0, \frac{1}{2})$. Let Λ be the pullback to $\mathbb{R}^d$ of the subgroup $\gamma(G)$ of $\mathbb{R}^d/\mathbb{Z}^d$, and write $\lambda_1, \ldots, \lambda_d$ for the successive minima of the cube $Q = (-1, 1)^d$ with respect to Λ. Define $r = \dim \operatorname{span}_{\mathbb{R}}(\Lambda \cap \rho\overline{Q})$. Then $B(\gamma, \rho)$ contains a proper coset progression $H + P$ of rank r and size at least $(\rho/r)^r \lambda_{r+1} \cdots \lambda_d |G|$. In particular, $H + P$ has rank at most d and size at least $(\rho/d)^d |G|$.*

Proposition 3.6.3 *Let G be a finite abelian group, let $d \in \mathbb{N}$, let $\gamma \in \widehat{G}^d$, and let $\rho \in (0, \frac{1}{6})$. Let Λ be the pullback to $\mathbb{R}^d$ of the subgroup $\gamma(G)$ of $\mathbb{R}^d/\mathbb{Z}^d$, and write $\lambda_1, \ldots, \lambda_d$ for the successive minima of the cube $Q = (-1, 1)^d$ with respect to Λ. Define $r = \dim \operatorname{span}_{\mathbb{R}}(\Lambda \cap \rho\overline{Q})$. Then we have*

$$|B(\gamma, \rho)| \leq (12d)^d \rho^r \lambda_{r+1} \cdots \lambda_d |G|.$$

Proof of Proposition 3.6.2 Let $v_1, \ldots, v_d$ be a directional basis for Λ with respect to Q, and for $i = 1, \ldots, r$ write $L_i = \rho/(r\lambda_i)$. Note that

$$P(v_1, \ldots, v_r; L_1, \ldots, L_r) \subset \rho\overline{Q} \cap \Lambda. \tag{3.6.2}$$

Set $H = \ker \gamma$, pick an arbitrary $x_i \in G$ such that $\gamma(x_i) \equiv v_i \pmod{\mathbb{Z}^d}$ for each $i \in [r]$, and set

$$P = P(x_1, \ldots, x_r; L_1, \ldots, L_r),$$

noting that $H + P \subset B(\gamma, \rho)$. We claim that $H + P$ is proper. Indeed, let $\ell_1, \ldots, \ell_r, \ell'_1, \ldots, \ell'_r$ with $|\ell_i| \leq L_i$ be such that

$$\ell_1 x_1 + \cdots + \ell_r x_r \in \ell'_1 x_1 + \cdots + \ell'_r x_r + H.$$

By (3.6.2) we then have

$$(\ell_1 - \ell'_1)v_1 + \cdots + (\ell_r - \ell'_r)v_r \in 2\rho\overline{Q} \cap \mathbb{Z}^d,$$

which by the linear independence of the v_i and the fact that $\rho < \frac{1}{2}$ implies that $\ell_i = \ell_i'$ for every i, and so $H + P$ is proper as claimed.

Now note that $\det(\Lambda) = |H|/|G|$ and $\mathrm{vol}(Q) = 2^d$, so that Minkowski's second theorem (Theorem 3.5.1) gives

$$\lambda_1 \cdots \lambda_d \leq \frac{|H|}{|G|}. \tag{3.6.3}$$

We therefore have

$$\begin{aligned} |H + P| &\geq L_1 \cdots L_r |H| && \text{(by properness)} \\ &= \frac{\rho^r |H|}{r^r \lambda_1 \cdots \lambda_r} \\ &\geq (\rho/r)^r \lambda_{r+1} \cdots \lambda_d |G| && \text{(by (3.6.3))}, \end{aligned}$$

and so $H + P$ is of the required size and the proof is complete. □

In proving Proposition 3.6.3 it will be convenient to define, for a given symmetric convex body $B \subset \mathbb{R}^d$, the norm $\|\cdot\|_B$ on $\mathbb{R}^d$ via $\|x\|_B = \inf\{\nu \geq 0 : x \in \nu B\}$. It turns out that $\|\cdot\|_B$ is indeed a norm. However, as we will not need this fact we leave it to the reader to prove it in Exercise 3.6, along with the converse statement that the unit ball of an arbitrary norm on $\mathbb{R}^d$ is a symmetric convex body.

Lemma 3.6.4 *Let B be a symmetric convex body in $\mathbb{R}^d$ and let Λ be a lattice in $\mathbb{R}^d$. Let $\lambda_1, \ldots, \lambda_d$ be the successive minima of B with respect to Λ, and define $r = \dim \mathrm{span}_{\mathbb{R}}(\Lambda \cap \overline{B})$. Then there exists a basis $x_1, \ldots, x_d$ for $\mathbb{R}^d$ with $x_i \in \Lambda$ for each i such that $1 < \|x_i\|_B \leq 2$ for $i = 1, \ldots, r$ and $\|x_i\|_B = \lambda_i$ for $i = r+1, \ldots, d$, and such that $\overline{B} \cap \langle x_1, \ldots, x_d \rangle = \{0\}$.*

Proof Let $v_1, \ldots, v_d$ be a directional basis for Λ with respect to B. For $i = d, \ldots, 1$ in turn, define

$$\alpha_i = \min\{\alpha \in \mathbb{N} : \|\alpha \alpha_{i+1} \cdots \alpha_d v_i\|_B > 1\},$$

noting that

$$\alpha_k \cdots \alpha_d > \lambda_k^{-1} \tag{3.6.4}$$

for every k. Set $x_i = \alpha_i \cdots \alpha_d v_i$ for each i, noting that $1 < \|x_i\|_B \leq 2$ for $i = 1, \ldots, r$ and $\|x_i\| = \lambda_i$ for $i = r + 1, \ldots, d$, as required.

It remains to show that $\overline{B} \cap \langle x_1, \ldots, x_d \rangle = \{0\}$. To that end, let $y \in \overline{B} \cap \langle x_1, \ldots, x_d \rangle$, and let k be minimal such that $y \in \mathrm{span}_{\mathbb{R}}(v_1, \ldots, v_k)$. This implies that $y \in \langle x_1, \ldots, x_k \rangle$, and hence $y \in \alpha_k \cdots \alpha_d \langle v_1, \ldots, v_k \rangle$.

It follows that

$$\frac{y}{\alpha_k \cdots \alpha_d} \in \left(\frac{1}{\alpha_k \cdots \alpha_d}\overline{B}\right) \cap \Lambda,$$

and hence by (3.6.4) that

$$\frac{y}{\alpha_k \cdots \alpha_d} \in \lambda_k B \cap \Lambda.$$

By the minimality of k and the definitions of λ_k and v_k it follows that $y = 0$. We therefore have $\overline{B} \cap \langle x_1, \dots, x_d \rangle = \{0\}$, as required. □

Proof of Proposition 3.6.3 Write $H = \ker \gamma$, and note that

$$|\rho\overline{Q} \cap \Lambda||H| = |B(\gamma, \rho)| \tag{3.6.5}$$

and

$$|\tfrac{1}{2}Q \cap \Lambda||H| \leq |G|. \tag{3.6.6}$$

Let $x_1, \dots, x_d \in \Lambda$ be the basis for $\mathbb{R}^d$ arising on applying Lemma 3.6.4 with $B = 2\rho Q$, noting that the successive minima of $2\rho Q$ with respect to Λ are $\lambda_i/2\rho$. Note in particular that

$$2\rho\overline{Q} \cap \langle x_1, \dots, x_d \rangle = \{0\}. \tag{3.6.7}$$

Since $\|\cdot\|_\infty = 2\rho\|\cdot\|_B$, we have $\|x_i\|_\infty \leq 4\rho$ for $i = 1, \dots, r$ and $\|x_i\|_\infty = \lambda_i$ for $i = r+1, \dots, d$. Defining $L_i = \lfloor 1/(12d\rho) \rfloor$ for $i = 1, \dots, r$, and $L_i = \lfloor 1/(3d\lambda_i) \rfloor$ for $i = r+1, \dots, d$, it follows that $P(x; L) \subset \frac{1}{3}Q$, and hence, since $\rho < \frac{1}{6}$, that

$$P(x; L) + \rho\overline{Q} \subset \tfrac{1}{2}Q. \tag{3.6.8}$$

Note that we have, slightly crudely,

$$|P(x; L)| \geq \frac{1}{(12d)^d \rho^r \lambda_{r+1} \cdots \lambda_d}. \tag{3.6.9}$$

Now, given two distinct elements $u, v \in P(x; L)$, we have $(u + \rho\overline{Q}) \cap (v + \rho\overline{Q}) = \varnothing$, since if $(u + \rho\overline{Q}) \cap (v + \rho\overline{Q}) \neq \varnothing$ for $u, v \in P(x; L)$ then $u - v \in 2\rho\overline{Q} \cap \langle x_1, \dots, x_d \rangle$ and then $u = v$ by (3.6.7). It therefore follows from (3.6.8) that $|\frac{1}{2}Q \cap \Lambda| \geq |P(x; L)||\rho\overline{Q} \cap \Lambda|$, and hence

$$\begin{aligned} |G| &\geq |P(x; L)||\rho\overline{Q} \cap \Lambda||H| && \text{(by (3.6.6))} \\ &\geq \frac{1}{(12d)^d \rho^r \lambda_{r+1} \cdots \lambda_d}|B(\gamma, \rho)| && \text{(by (3.6.5) and (3.6.9))}, \end{aligned}$$

and the proposition is proved. □

Exercises

3.1 Show that a coset progression of rank r in an arbitrary abelian group is a Freiman-homomorphic image of a Bohr set of rank r in some finite abelian group.

3.2 It follows from Proposition 3.6.2 that a Bohr set $B(\gamma, \rho)$ of rank d inside a finite abelian group G satisfies $|B(\gamma, \rho)| \geq (\rho/d)^d |G|$. Prove directly that in fact $|B(\gamma, \rho)| \geq \rho^d |G|$.

3.3 Let G be an abelian group, let $\pi : \mathbb{Z}^d \to G$ be a homomorphism, and let $B \subset \mathbb{R}^d$ be a symmetric convex body. Show that the set $\pi(B \cap \mathbb{Z}^d)$ is a K-approximate group for some K depending only on d. Noting that a progression is a special case of such a set in which B is a cuboid, formulate a similar generalisation of a Bohr set of rank d, and prove that it is a K-approximate group with K depending only on d.

3.4 Show that two bases $x_1, \ldots, x_d$ and $y_1, \ldots, y_d$ of $\mathbb{R}^d$ generate the same lattice Λ if and only if there exists an $n \times n$ matrix $A = (a_{ij})$ with integer entries and $\det(A) = \pm 1$ such that $y_i = \sum_{j=1}^{n} a_{ij} x_j$ for every i. Use this to give an alternative proof of the fact that $\det(\Lambda)$ is well defined.

3.5 Give an example, for some d, of a symmetric convex body $B \subset \mathbb{R}^d$ and a lattice $\Lambda \subset \mathbb{R}^d$ such that whenever $v_1, \ldots, v_d$ is a directional basis for Λ with respect to B we have $\langle v_1, \ldots, v_d \rangle \neq \Lambda$. What is the smallest d for which this is possible?

3.6 Show that if $B \subset \mathbb{R}^d$ is a symmetric convex body then $\|\cdot\|_B$ is a norm. Conversely, show that if $\|\cdot\|$ is an arbitrary norm on $\mathbb{R}^d$ then there exists a symmetric convex body B such that $\|\cdot\| = \|\cdot\|_B$.

4
Small Doubling in Abelian Groups

4.1 Introduction

In the previous chapter we identified some examples of sets of small doubling in abelian groups, and showed that each of them could be realised as a dense subset of some coset progression. It turns out that this is no coincidence: the following remarkable theorem shows that in an abelian group every set of small doubling is a dense subset of some coset progression.

Theorem 4.1.1 (Freiman–Green–Ruzsa theorem; simple form) *Let A be a finite subset of an abelian group such that $|A + A| \leq K|A|$. Then there exists a coset progression $H + P$ of rank at most $r(K)$ and size at most $c(K)|A|$ such that $A \subset H + P$.*

Theorem 4.1.1 was originally proved by Freiman [26] in the case of torsion-free G, and then generalised to arbitrary abelian G by Green and Ruzsa [35], building on earlier work of Ruzsa [55]. In this chapter we follow the Green–Ruzsa argument to prove the following slightly refined statement.

Theorem 4.1.2 (Freiman–Green–Ruzsa theorem; detailed form) *Let A be a finite subset of an abelian group such that $|A + A| \leq K|A|$. Then there exist a finite subgroup $H \subset 2A - 2A$, a non-negative integer $r \ll K^{O(1)}$, elements $x_1, \ldots, x_r \in 2A - 2A$, and natural numbers $L_1, \ldots, L_r$ such that the progression $P = P(x_1, \ldots, x_r; L_1, \ldots, L_r)$ satisfies*

$$A \subset H + P \subset O(K^{O(1)})(A - A).$$

Remark Note that by Theorem 2.3.1 the coset progression $H+P$ given by Theorem 4.1.2 satisfies $|H+P| \leq \exp(O(K^{O(1)}))|A|$, so Theorem 4.1.2 implies Theorem 4.1.1 with $r(K) \ll K^{O(1)}$ and $c(K) \leq \exp(O(K^{O(1)}))$.

The details of the proof are rather simpler in the case $A \subset \mathbb{Z}$, and in Exercises 4.1–4.3 we guide the reader through this simpler argument.

The bounds we obtain in Theorem 4.1.2 can be computed explicitly without difficulty, and indeed we keep track of them throughout most of the argument. The form $O(K^{O(1)})$ of the bounds is the same as obtained by Green and Ruzsa, although we opt to make the argument shorter or simpler in a few places at the expense of losing some tightness, resulting in slightly worse implied constants. We discuss the bounds in Theorem 4.1.2 in more detail in Remark 4.1.8 at the end of this section.

We saw in the proof of Theorem 3.1.6 at the beginning of Section 3.6 that, slightly counterintuitively, if one wishes to place a given set efficiently inside a coset progression then a good place to start is to locate a large coset progression inside that set. We adopt a similar approach in our proof of Theorem 4.1.2, and indeed at the heart of our proof is the following result.

Theorem 4.1.3 (Green–Ruzsa) *Let A be a finite subset of an abelian group such that $|A+A| \leq K|A|$. Then there exists a finite subgroup H, and a progression P of rank $O(K^{O(1)})$ such that $H+P \subset 2A-2A$ and $|H+P| \geq \exp(-O(K^{O(1)}))|A|$.*

We now give a brief overview of the proof of Theorems 4.1.2 and 4.1.3. One setting in which the proof of Theorem 4.1.3 turns out not to be too difficult is when A is a dense subset of a finite abelian group. In that setting we have the following special case of Theorem 4.1.3, which we prove in Section 4.5.

Proposition 4.1.4 *Let G be a finite group, let $A \subset G$ be a subset of size $\alpha|G|$, and suppose that $|A+A| \leq K|A|$. Then $2A-2A$ contains a proper coset progression $H+P$ of rank at most $4K^2/\alpha$ and size at least $(\alpha/24K^2)^{4K^2/\alpha}|G|$.*

Of course, for any fixed α, in the setting of Proposition 4.1.4 we trivially have the conclusion of Theorem 4.1.2; indeed, G itself is a coset progression of rank 0 and size $\alpha^{-1}|A|$ containing A. Nonetheless, it turns out that we can also force Proposition 4.1.4 to apply in the general case, where Theorem 4.1.2 is far from obvious, thanks to the following result, which we prove in Section 4.4.

Proposition 4.1.5 (Green–Ruzsa [35]) *Let A be a finite subset of an abelian group and suppose that $|A+A| \leq K$. Let $s \in \mathbb{N}$ with $s \geq 2$. Then A has a Freiman s-model in a group G satisfying*

$$|G| \leq 4K^{20}(200sK^2)^{2K^{16}}|A|.$$

The precise value of the bound $4K^{20}(200sK^2)^{2K^{16}}$ coming from Proposition 4.1.5 is not too important, and is weaker than the bound obtained by Green and Ruzsa; the main thing to note is that it is bounded by a function of the form $\exp(O_s(K^{O_s(1)}))$.

Combining Propositions 4.1.4 and 4.1.5 allows us to locate a large coset progression inside an arbitrary set of small doubling in an abelian group. Indeed, given A satisfying $|A+A| \leq K|A|$, Proposition 4.1.5 implies that there is a Freiman 8-isomorphism from A to a subset A' of density at least $\exp(-O(K^{O(1)}))$ in some finite abelian group G', Proposition 4.1.4 implies that $2A' - 2A'$ contains a proper coset progression $H' + P'$ of rank at most $\exp(O(K^{O(1)}))$ and size at least

$$\exp(-\exp(O(K^{O(1)})))|A'|,$$

and then Lemmas 2.7.3 (ii) and 3.2.2 allow us to pull that coset progression back to a proper coset progression $H + P \subset 2A - 2A$ of rank at most $\exp(O(K^{O(1)}))$ and size at least $\exp(-\exp(O(K^{O(1)})))|A|$.

This is already enough to prove a version of Theorem 4.1.2, using Ruzsa's covering lemma (Lemma 2.4.4) in the same way as in the proof of Theorem 3.1.6. However, such an argument gives only bounds of the form $\exp(\exp(O(K^{O(1)})))$ in Theorem 4.1.2, which is significantly worse than what we have claimed. Chang [22] found two ways to make this argument more efficient, each of which saves an exponential from the bound. One of her refinements was to the covering part of the argument, and gives rise to the following result, which we prove in Section 4.7.

Proposition 4.1.6 (Chang) *Let A be a finite subset of an abelian group such that $|A+A| \leq K|A|$, and suppose that $P \subset mA - nA$ is another subset satisfying $|P| \geq |A|/C$. Then there is a progression Q of rank at most*

$$2K\left(1 + \frac{(m+n+1)\log K + \log C}{\log 2}\right)$$

such that

$$A \subset P - P + Q$$
$$\subset \left(2K\left(1 + \frac{(m+n+1)\log K + \log C}{\log 2}\right) + m + n\right)(A - A).$$

The key point to note about Proposition 4.1.6 is the fact that the dependence of the bounds on C is logarithmic. If $P \subset 2A - 2A$ is the progression coming from Propositions 4.1.4 and 4.1.5, we have $C = \exp(\exp(O(K^{O(1)})))$, and so this logarithmic dependence results in the removal of one exponential, and it turns out that this improves the bounds in Theorem 4.1.2 to $\exp(O(K^{O(1)}))$.

In another refinement, Chang was able to remove a certain amount of redundancy from the progression coming from Proposition 4.1.4, making its rank merely logarithmic in α^{-1}, rather than linear in α^{-1}, as follows.

Proposition 4.1.7 *Let G be a finite abelian group, let $A \subset G$ be a subset of size $\alpha|G|$, and suppose that $|A + A| \leq K|A|$. Then $2A - 2A$ contains a proper coset progression $H + P$ of rank at most $16K^2 \log \alpha^{-1}$ and size at least $(1/1536K^4 \log^2 \alpha^{-1})^{16K^2 \log \alpha^{-1}}|G|$.*

Again, this logarithmic dependence results in the removal of an exponential from the bounds in Theorem 4.1.2 and, as we will see shortly, gives the claimed bounds in Theorem 4.1.2. We give an overview of the refinement from Proposition 4.1.4 to Proposition 4.1.7 in Section 4.6. However, the details are somewhat more technical than those of the rest of the argument, so in order to keep focused on the main ideas we defer the full proof until the appendix to this chapter, Section 4.A. A first-time reader might reasonably at that point accept the worse bounds obtained by using Proposition 4.1.4 and move on to the next chapter.

We close this introduction by combining the results summarised above into proofs of Theorems 4.1.2 and 4.1.3.

Proof of Theorem 4.1.3 By Proposition 4.1.5 there exists a subset A' inside a finite abelian group G satisfying $|A'| \geq \exp(-O(K^{O(1)}))|G|$ and a Freiman 8-isomorphism $\pi : A' \to A$. Proposition 4.1.7 implies that $2A' - 2A'$ contains a proper coset progression $H_0 + P_0$ of rank at most $O(K^{O(1)})$ and size at least $\exp(-O(K^{O(1)}))|A|$. Lemma 2.7.3 (ii) implies that π induces a centred Freiman 2-isomorphism $\pi' : 2A' - 2A' \to 2A - 2A$, and then Lemma 3.2.2 implies that $\pi'(H_0 + P_0)$ is a coset progression of rank at most $O(K^{O(1)})$ and size at least $\exp(-O(K^{O(1)}))|A|$ in $2A - 2A$, as required. □

Proof of Theorem 4.1.2 It follows from Theorem 4.1.3 that there is a coset progression $H + P$ of rank at most $O(K^{O(1)})$ and size at least $\exp(-O(K^{O(1)}))|A|$ in $2A - 2A$. Proposition 4.1.6 then implies that there is a progression Q of rank at most $O(K^{O(1)})$ such that $A \subset H + P - P + Q \subset O(K^{O(1)})(A - A)$. The set $H + P - P + Q$ is a coset progression of rank at most $O(K^{O(1)})$, and so this completes the proof. □

Remark 4.1.8 (bounds in Theorem 4.1.1) The best bound one could hope to achieve on the quantity $r(K)$ appearing in Theorem 4.1.1 is essentially $2K$, since the set $e_1, \ldots, e_n$ of standard generators of $\mathbb{Z}^n$ has doubling roughly $\frac{1}{2}n$ and does not lie in any subgroup of rank less than n, so in particular does not lie in any progression of rank less than n. The same example also shows that the best bound one could hope to achieve on $c(K)$ is exponential in K.

The best bounds so far actually achieved in Theorem 4.1.1 differ depending on whether one wishes to prioritise $r(K)$ or $c(K)$. Sanders [59, Theorem 11.4] has shown that one may take $r(K) \leq O(K \log^{O(1)} 2K)$ and $c(K) \leq \exp(O(K \log^{O(1)} 2K))$ (see Exercise 4.4 for slightly more detail). On the other hand, Cwalina and Schoen [24, Theorem 12] have shown that one can essentially take $r(K) \leq (2 + o(1))K$ and $c(K) \leq \exp(O(K^2 \log K))$ (so $r(K)$ slightly better and $c(K)$ slightly worse than in the Sanders result).

One can, however, make $c(K)$ merely polynomial in K if one is prepared to relax slightly the qualitative conclusion of Theorem 4.1.1 and show that A is contained in a few translates of a coset progression, rather than a coset progression itself. Put another way, rather than requiring that A be a dense subset of a coset progression $H + P$, one needs to be prepared to require that A be a dense subset of the sum $X + H + P$ of a 'small' set X and a coset progression $H + P$. For example, in Exercise 4.4 we invite the reader to deduce from part of Sanders's work that if A is a subset of an abelian group of doubling at most K then there exists a subset $X \subset A$ of size at most $\exp(O(\log^{O(1)} 2K))$ and a coset progression $H + P \subset 4A - 4A$ of rank at most $O(\log^{O(1)} 2K)$ such that $A \subset X + H + P$. In particular, Theorem 2.3.1 then implies that $|X + H + P| \leq K^9|A|$.

A famous conjecture called the *polynomial Freiman–Ruzsa conjecture* posits that in addition to the density of A in $X + H + P$ being polynomial in K, the size of X should be polynomial in K and the rank of P should be logarithmic in K (we leave it to the reader to check that these bounds would be optimal). We caution, however, that Lovett and Regev [43] have shown that such a statement requires a formulation of the notion of progression along the lines discussed in Exercise 3.3, which is more general than Definition 3.1.1.

4.2 Fourier Analysis

An extremely useful tool for understanding additive structure in abelian groups that is central to the arguments of this chapter is the theory of *Fourier analysis*. Fourier analysis is essentially the process of expressing a function $f : G \to \mathbb{C}$ with respect to a certain orthonormal basis for $\mathbb{C}^G$. This can be done for any abelian group G, but for our purposes it will be sufficient to restrict attention to the case in which G is finite. In that case we define an inner product on the functions $G \to \mathbb{C}$ via $\langle f, g \rangle = \mathbb{E}_{x \in G} f(x)\overline{g(x)}$. We then use the group $\widehat{G}$ of homomorphisms $G \to \mathbb{T}$ to define the orthonormal basis with respect to which we seek to express f.

It follows from Theorem 1.5.1 that a finite abelian group G is of the form

$$G = \mathbb{Z}/n_1\mathbb{Z} \oplus \cdots \oplus \mathbb{Z}/n_r\mathbb{Z}$$

for some integers $n_i \geq 2$, and having written G in this form an element $\gamma \in \widehat{G}$ can be given in the form

$$\gamma = (\gamma_1, \ldots, \gamma_r) \in \mathbb{Z}/n_1\mathbb{Z} \oplus \cdots \oplus \mathbb{Z}/n_r\mathbb{Z}.$$

Indeed, defining an inclusion homomorphism $\psi_n : \mathbb{Z}/n\mathbb{Z} \to \mathbb{T}$ via $\psi_n(k) = \frac{k}{n}$, the map $\gamma = (\gamma_1, \ldots, \gamma_r)$ is given by

$$\gamma(x_1, \ldots, x_r) = \psi_{n_1}(\gamma_1 x_1) + \cdots + \psi_{n_r}(\gamma_r x_r). \tag{4.2.1}$$

In particular, this characterisation of $\widehat{G}$ makes it clear that

$$\widehat{G} \cong G. \tag{4.2.2}$$

To convert the elements of $\widehat{G}$ into functions $G \to \mathbb{C}$, we write $S^1 = \{z \in \mathbb{C} : |z| = 1\}$ and define a map $e : \mathbb{T} \to S^1$ via $e(\theta) = e^{2\pi i \theta}$. Defining $\zeta_\gamma : G \to \mathbb{C}$ via $\zeta_\gamma(x) = e(\gamma(x))$ then defines a homomorphism of G into S^1; we call such a homomorphism a *character* of G.

It turns out that the characters of G form an orthonormal basis of $\mathbb{C}^G$ with respect to the inner product we defined above, which is to say that they span $\mathbb{C}^G$ as a vector space and satisfy

$$\mathbb{E}_{x \in G} e(\gamma(x) - \gamma'(x)) = \begin{cases} 1 \text{ if } \gamma = \gamma', \\ 0 \text{ if } \gamma \neq \gamma'. \end{cases} \tag{4.2.3}$$

Indeed, if $\gamma = \gamma'$ then this certainly has value 1. On the other hand, if $\gamma \neq \gamma'$ then we may pick some $x_0 \in G$ for which $\gamma(x_0) \neq \gamma'(x_0)$ and write

$$\begin{aligned} \mathbb{E}_{x \in G} e(\gamma(x) - \gamma'(x)) &= \mathbb{E}_{x \in G} e(\gamma(x + x_0) - \gamma'(x + x_0)) \\ &= e(\gamma(x_0) - \gamma'(x_0)) \mathbb{E}_{x \in G} e(\gamma(x) - \gamma'(x)). \end{aligned}$$

Since $e(\gamma(x_0) - \gamma'(x_0)) \neq 1$ by the choice of x_0, it follows that

$$\mathbb{E}_{x \in G} e(\gamma(x) - \gamma'(x)) = 0.$$

This shows that the characters are orthonormal; the fact that they are a basis then follows from (4.2.2), which implies in particular that their span has dimension $|G|$ and so must be the whole of $\mathbb{C}^G$.

Fourier analysis is the process of expressing $f : G \to \mathbb{C}$ with respect to this basis of characters, which is to say identifying the function $\widehat{f} : \widehat{G} \to \mathbb{C}$ that satisfies

$$f(x) = \sum_{\gamma \in \widehat{G}} \widehat{f}(\gamma) e(-\gamma(x)). \tag{4.2.4}$$

The function $\widehat{f}$ is called the *Fourier transform* of f, and the individual values it takes, $\widehat{f}(\gamma)$, are called *Fourier coefficients*. It is not hard to check that the Fourier coefficient $\widehat{f}(\gamma)$ is given by

$$\widehat{f}(\gamma) = \mathbb{E}_{x \in G} f(x) e(\gamma(x)). \tag{4.2.5}$$

Indeed, this follows easily from the fact that

$$\sum_{\gamma \in \widehat{G}} e(\gamma(x)) = \begin{cases} |G| \text{ if } x = 0, \\ 0 \text{ if } x \neq 0, \end{cases}$$

which is proved in a similar way to (4.2.3), but using the fact that for a non-zero $x \in G$ there exists $\gamma_0 \in \widehat{G}$ such that $\gamma_0(x) \neq 0$.

Another useful identity to note is *Plancherel's theorem*, which states that

$$\mathbb{E}_{x \in G} f(x) \overline{g(x)} = \sum_{\gamma \in \widehat{G}} \widehat{f}(\gamma) \overline{\widehat{g}(\gamma)} \tag{4.2.6}$$

and may be verified easily using (4.2.3). In the special case of $f = g$ Plancherel's theorem reduces to *Parseval's identity*, which states that

$$\mathbb{E}_{x \in G}|f(x)|^2 = \sum_{\gamma \in \widehat{G}} |\widehat{f}(\gamma)|^2. \tag{4.2.7}$$

In this book we are interested in subsets A of a group G, rather than functions on G, but it is a straightforward matter to convert A into a function by identifying it with its indicator function 1_A, which we may then study via its Fourier transform $\widehat{1}_A$. To emphasise this identification, we will in general refer to $\widehat{1}_A$ as simply the 'Fourier transform of A'.

Since a set A can be completely recovered from its Fourier transform via the formula (4.2.4), the structure of A is in principle completely encoded in the Fourier transform. One very easy example of this is the fact that

$$\widehat{1}_A(0) = \frac{|A|}{|G|}, \tag{4.2.8}$$

and so the Fourier coefficient at zero gives the density of A in G. The idea of many of the arguments of this chapter is to exploit more general instances of this phenomenon. The general scheme will be to use small doubling of A or some similar hypothesis to extract information about the Fourier transform of A, and then to decode that information in order to say something explicit about the structure of A.

One example of the kind of information we can extract about the Fourier transform of a set of small doubling is the following.

Lemma 4.2.1 (Green–Ruzsa [34, Lemma 4.1]) *Let G be a finite abelian group, let $A \subset G$, and write $\alpha = |A|/|G|$. Suppose that $|A + A| \leq K|A|$ and $\alpha \leq (2 \cdot 3^{K^4} K^2)^{-1}$. Then there exists a non-zero $\gamma \in \widehat{G}$ such that*

$$|\widehat{1}_A(\gamma)| \geq \left(1 - (2K^2\alpha)^{1/K^4} \log \alpha^{-1}\right) \frac{|A|}{|G|}.$$

For small α, the conclusion of Lemma 4.2.1 can be interpreted as saying that $|\widehat{1}_A(\gamma)|$ is 'large', since it is immediate from (4.2.5) that $|A|/|G|$ is the maximum possible absolute value for a Fourier coefficient of A. This can then indeed be converted into more explicit information about the structure of A, specifically that A is concentrated in a few cosets of $\ker \gamma$, as follows.

Lemma 4.2.2 (Green–Ruzsa [35, Lemma 2.4]) *Let $\eta \in (0,1)$. Let G be a finite abelian group, let $A \subset G$, and let $\gamma \in \widehat{G}$. Suppose that*

$$|\widehat{1}_A(\gamma)| = (1-\eta)\frac{|A|}{|G|}. \tag{4.2.9}$$

Then given $\delta \in (0,1]$ there exists an interval $U \subset \mathbb{T}$ of width at most δ such that

$$|A \cap \gamma^{-1}(U)| \geq (1 - \eta\delta^{-2})|A|.$$

Remark The reader is invited to show in Exercise 4.7 that the conclusion of Lemma 4.2.2 cannot be strengthened to say that $A \subset \gamma^{-1}(U)$.

The proof of Lemma 4.2.1 requires the additional concept of the *convolution* of two functions; we introduce this concept and prove Lemma 4.2.1 in the next section.

The intuition of Lemma 4.2.2 is very simple: (4.2.5) may in this case be rewritten as $\widehat{1}_A(\gamma) = \frac{1}{|G|}\sum_{x\in A} e(\gamma(x))$, and so if $|\widehat{1}_A(\gamma)|$ is large this means that the unit complex numbers $e(\gamma(x))$ must be biased around the direction of $\widehat{1}_A(\gamma)$. We now present the formal details.

Proof of Lemma 4.2.2 Write λ for the argument of the complex number $\widehat{1}_A(\gamma)$, so that

$$\begin{aligned} |\widehat{1}_A(\gamma)| &= e(-\lambda)\widehat{1}_A(\gamma) \\ &= \mathbb{E}_{x\in G}\widehat{1}_A(x)e(\gamma(x)-\lambda) \\ &= \frac{1}{|G|}\sum_{x\in A} e(\gamma(x)-\lambda) \\ &= \frac{1}{|G|}\sum_{x\in A} \cos 2\pi\|\gamma(x)-\lambda\|_{\mathbb{T}}, \end{aligned}$$

the last equality coming from the fact that the sum is real. The assumption (4.2.9) then implies that

$$(1-\eta)|A| = \sum_{x\in A} \cos 2\pi\|\gamma(x)-\lambda\|_{\mathbb{T}}. \tag{4.2.10}$$

Now when $\|\gamma(x)-\lambda\|_{\mathbb{T}} \geq \delta/2$ we have $\cos 2\pi\|\gamma(x)-\lambda\|_{\mathbb{T}} \leq \cos\pi\delta \leq 1-\delta^2$, whilst for every x we have $\cos 2\pi\|\gamma(x)-\lambda\|_{\mathbb{T}} \leq 1$. Writing $U = [\lambda-\delta/2, \lambda+\delta/2]$, it therefore follows from (4.2.10) that

$$(1-\eta)|A| \leq |A\cap\gamma^{-1}(U)| + (1-\delta^2)|A\setminus\gamma^{-1}(U)|.$$

Writing $\rho = |A\cap\gamma^{-1}(U)|/|A|$ and dividing through by $|A|$, this implies that $(1-\eta) \leq \rho + (1-\delta^2)(1-\rho)$, from which it easily follows that $\rho \geq 1 - \eta\delta^{-2}$, as required. □

4.3 Convolutions and Fourier Analysis of Sets of Small Doubling

The main aim of this section is to prove Lemma 4.2.1. First, however, we introduce a quite general Fourier-analytic tool that we use in the proof, namely the *convolution* of two functions.

Given two functions $f, g : G \to \mathbb{C}$, the *convolution* $f * g : G \to \mathbb{C}$ of f and g is defined via

$$f * g(x) = \mathbb{E}_{y \in G} f(y) g(x-y).$$

We abbreviate higher convolution powers with exponents in brackets, so that

$$f^{(n)} = \underbrace{f * \cdots * f}_{n \text{ terms}}.$$

It is a straightforward exercise to verify that convolution is associative, a fact that we use implicitly throughout this chapter without further mention.

We leave to the reader the simple exercise of verifying that

$$\mathbb{E}(f * g) = (\mathbb{E}f)(\mathbb{E}g) \tag{4.3.1}$$

and

$$\widehat{f * g} = \widehat{f}\widehat{g} \tag{4.3.2}$$

for every $g, f : G \to \mathbb{C}$. Since $\widehat{1}_{-A} = \overline{\widehat{1}_A}$, (4.3.2) implies in particular that

$$\widehat{1_A * 1_{-A}}(\gamma) = |\widehat{1}_A(\gamma)|^2. \tag{4.3.3}$$

One reason that convolution arises in the study of sum sets is that if $A_1, \dots, A_k$ are sets in an abelian group then the support $\operatorname{supp}(1_{A_1} * \cdots * 1_{A_k})$ of the convolution $1_{A_1} * \cdots * 1_{A_k}$ satisfies

$$\operatorname{supp}(1_{A_1} * \cdots * 1_{A_k}) = A_1 + \cdots + A_k. \tag{4.3.4}$$

Indeed, $1_{A_1} * \cdots * 1_{A_k}(x)$ is equal to

$$\mathbb{E}_{y_1,\dots,y_{k-1} \in G} 1_{A_1}(y_1) 1_{A_2}(y_2 - y_1) \cdots 1_{A_{k-1}}(y_{k-2} - y_{k-1}) 1_{A_k}(x - y_{k-1}),$$

and so in fact we have

$$1_{A_1} * \cdots * 1_{A_k}(x) = \frac{|\{(a_1, \ldots, a_k) \in A_1 \times \cdots \times A_k : a_1 + \cdots + a_k = x\}|}{|G|^{k-1}}, \tag{4.3.5}$$

and hence $1_{A_1} * \cdots * 1_{A_k}(x)$ gives a normalised count of the number of ways x can be written as a sum of elements from $A_1, \ldots, A_k$. To see why $1/|G|^{k-1}$ is a natural normalisation factor, note that

$$|G|^{k-1} = |\{(g_1, \ldots, g_k) \in G^k : g_1 + \cdots + g_k = x\}|;$$

thus, phrased in yet another way, $1_{A_1} * \cdots * 1_{A_k}(x)$ is the proportion of those sums $g_1 + \cdots + g_k = x$ with $g_i \in G$ for which $g_i \in A_i$ for every i.

We prove Lemma 4.2.1 following Green and Ruzsa [34, Lemma 4.1], starting with the following lemma, which bounds sums of powers of the Fourier coefficients of A in terms of the growth of the iterated sum sets of A.

Lemma 4.3.1 *Let G be a finite abelian group, let $A \subset G$, and write $\alpha = |A|/|G|$. Then*

$$\sum_{\gamma \in \widehat{G}} |\widehat{1}_A(\gamma)|^{2m} \geq \alpha^{2m-1} \frac{|A|}{|mA|}.$$

The idea is then to use the bounds on the sizes of iterated sum sets given by Corollary 2.4.3 to show that the terms $|\widehat{1}_A(\gamma)|^{2m}$ must be large on average, and then to show that this must mean that $|\widehat{1}_A(\gamma)|$ is large for at least one non-zero $\gamma \in \widehat{G}$.

Proof of Lemma 4.3.1 By (4.3.4) we have

$$1_A^{(m)}(x) = 1_A^{(m)}(x) 1_{mA}(x),$$

and so the Cauchy–Schwarz inequality (1.5.1) gives

$$|\mathbb{E}_{x \in G}\, 1_A^{(m)}(x)|^2 \leq \mathbb{E}_{x \in G}\, 1_A^{(m)}(x)^2\, \mathbb{E}_{x \in G}\, 1_{mA}(x)^2.$$

Since $1_{mA}(x)^2 = 1_{mA}(x)$, and since $\mathbb{E}_{x \in G}\, 1_A^{(m)}(x) = \alpha^m$ by (4.3.1), this implies that

$$\begin{aligned} \alpha^{2m} &\leq \mathbb{E}_{x \in G}\, 1_A^{(m)}(x)^2 \frac{|mA|}{|G|} \\ &= \alpha\, \mathbb{E}_{x \in G}\, 1_A^{(m)}(x)^2 \frac{|mA|}{|A|}. \end{aligned} \tag{4.3.6}$$

We therefore have

$$\begin{aligned}\sum_{\gamma\in\widehat{G}} |\widehat{1}_A(\gamma)|^{2m} &= \sum_{\gamma\in\widehat{G}} |\widehat{1_A^{(m)}}(\gamma)|^2 && \text{(by (4.3.2))}\\ &= \mathbb{E}_{x\in G}\, 1_A^{(m)}(x)^2 && \text{(by (4.2.7))}\\ &\geq \alpha^{2m-1}\frac{|A|}{|mA|} && \text{(by (4.3.6))},\end{aligned}$$

as required. □

Proof of Lemma 4.2.1 Following Green and Ruzsa [34, Lemma 4.1], we have $\widehat{1}_A(0) = \alpha$ by (4.2.8), and so subtracting the contribution of the $\gamma = 0$ term from the sum $\sum_{\gamma\in\widehat{G}} |\widehat{1}_A(\gamma)|^{2m+2}$ and applying Lemma 4.3.1 gives

$$\sum_{\gamma\neq 0} |\widehat{1}_A(\gamma)|^{2m+2} \geq \alpha^{2m+1}\left(\frac{|A|}{|(m+1)A|} - \alpha\right).$$

This implies in particular that

$$\max_{\gamma\neq 0} |\widehat{1}_A(\gamma)|^{2m} \sum_{\gamma\neq 0} |\widehat{1}_A(\gamma)|^2 \geq \alpha^{2m+1}\left(\frac{|A|}{|(m+1)A|} - \alpha\right).$$

Since $\sum_{\gamma\neq 0} |\widehat{1}_A(\gamma)|^2 \leq \alpha$ by Parseval's identity (4.2.7), we may conclude that

$$\max_{\gamma\neq 0} |\widehat{1}_A(\gamma)|^{2m} \geq \alpha^{2m}\left(\frac{|A|}{|(m+1)A|} - \alpha\right),$$

and then applying Corollary 2.4.3 and taking $(2m)$th roots gives

$$\max_{\gamma\neq 0} |\widehat{1}_A(\gamma)| \geq \left(\frac{1}{K^2(m+1)^{K^4}} - \alpha\right)^{1/2m}\frac{|A|}{|G|}.$$

Setting $m = \lfloor (2K^2\alpha)^{-1/K^4}\rfloor - 1$, this implies that

$$\begin{aligned}\max_{\gamma\neq 0} |\widehat{1}_A(\gamma)| &\geq \alpha^{1/2m}\frac{|A|}{|G|}\\ &= e^{-\frac{1}{2m}\log\alpha^{-1}}\frac{|A|}{|G|}\\ &> \left(1 - \frac{\log\alpha^{-1}}{2m}\right)\frac{|A|}{|G|}.\end{aligned} \tag{4.3.7}$$

The assumption that $\alpha \leq (2\cdot 3^{K^4}K^2)^{-1}$ implies that for this choice of m we have

$$m \geq \frac{1}{2(2K^2\alpha)^{1/K^4}},$$

and so (4.3.7) implies that

$$\max_{\gamma \neq 0} |\widehat{1}_A(\gamma)| \geq \left(1 - (2K^2\alpha)^{1/K^4} \log \alpha^{-1}\right) \frac{|A|}{|G|},$$

and the lemma is satisfied. □

4.4 Dense Models for Abelian Sets of Small Doubling

In this section we prove Proposition 4.1.5. Before doing so, it is worth remarking that the generality of Proposition 4.1.5 makes it considerably harder to prove. In Exercises 4.1 and 4.5 we guide the reader to analogous results with better bounds and considerably simpler proofs when A is a subset of either $\mathbb{Z}$ or $\mathbb{F}_2^n$.

We start with the straightforward observation that A at least has a model in *some* finite group.

Lemma 4.4.1 *Let G be an abelian group and let $A \subset G$ be a finite set. Then for every s there exists a Freiman s-model of A inside some finite abelian group.*

Proof We may assume that G is generated by A. This implies in particular that G is finitely generated, and hence isomorphic to $H \times \mathbb{Z}^d$ for some finite abelian group H and some $d \in \mathbb{N}$. Since A is finite, we may therefore view it as a subset of $H \times [-n, n]^d$ for some $n \in \mathbb{N}$. The quotient homomorphism $\pi : H \times \mathbb{Z}^d \to H \times \mathbb{Z}^d / 4sn\mathbb{Z}^d$ therefore restricts to a Freiman s-isomorphism of A into $H \times \mathbb{Z}^d / 4sn\mathbb{Z}^d$, and the result follows. □

Lemma 4.4.1 means that in proving Proposition 4.1.5 we may assume to begin with that A is a subset of a finite group G. The idea of the proof is then to show that A can be modelled in a sequence of smaller and smaller groups, until we finally reach a model in a group that is small enough for our purposes.

One situation in which it is trivial to pass to a model in a smaller group is if A is contained in a coset of some proper subgroup H of G; indeed, if $A \subset H + z$ then $A - z$ is an s-model of A in H by Lemma 2.7.3 (iii). Another easy case is if $A \subset \mathbb{Z}/n\mathbb{Z}$ and A is contained in a small interval, say $[0, n/4s)$: in that case, we may simply regard A as lying inside

$\mathbb{Z}/(n-1)\mathbb{Z}$. It is not too difficult to combine these two cases and arrive at the following result.

Lemma 4.4.2 *Let G be a finite abelian group, let $A \subset G$, and let $s, n \geq 2$. Suppose that $\varphi : G \to \mathbb{Z}/n\mathbb{Z}$ is a homomorphism and there exists $x \in \mathbb{Z}/nZ$ and $d < n/4s$ such that $\varphi(A) \subset [x, x+d]$. Then A has a Freiman k-model in* $(\ker \varphi) \times (\mathbb{Z}/(n-1)\mathbb{Z})$.

Proof We follow Green and Ruzsa [35, Proposition 1.2]. Upon translating A if necessary we may assume that $x = 0$. Pick $z \in \varphi^{-1}(1)$ so that $G = \ker \varphi + z$, and define maps $h : G \to \ker \varphi$ and $\lambda : G \to [0, n-1]$ in such a way that $g = h(g) + \lambda(g)z$ for every $g \in G$. We then define $\psi : A \to (\ker \varphi) \times (\mathbb{Z}/(n-1)\mathbb{Z})$ by setting

$$\psi(g) = (h(g), \lambda(g)).$$

The fact that $\varphi(A) \subset [0, n-2]$ implies that ψ is injective. We claim also that ψ is a Freiman s-isomorphism. To see that ψ is a Freiman s-homomorphism, suppose that $a_1, \ldots, a_s, a'_1, \ldots, a'_s \in A$ satisfy

$$a_1 + \cdots + a_s = a'_1 + \cdots + a'_s,$$

and note that this implies that $\varphi(a_1) + \cdots + \varphi(a_s) = \varphi(a'_1) + \cdots + \varphi(a'_s)$, and hence, since $\varphi(a_i), \varphi(a'_i) \in [0, n/4s)$ for each i, that

$$\lambda(a_1) + \cdots + \lambda(a_s) = \lambda(a'_1) + \cdots + \lambda(a'_s).$$

This in turn implies that $h(a_1) + \cdots + h(a_s) = h(a'_1) + \cdots + h(a'_s)$, and hence that

$$\psi(a_1) + \cdots + \psi(a_s) = \psi(a'_1) + \cdots + \psi(a'_s).$$

The proof that ψ^{-1} is a Freiman s-homomorphism on $\psi(A)$ is almost identical. □

We know from Lemmas 4.2.1 and 4.2.2 that if A has small doubling then a large proportion of it is contained in the preimage under a homomorphism of some small interval, similar to the hypothesis of Lemma 4.4.2. However, the hypothesis of Lemma 4.4.2 requires the whole of A to be contained in such a set, and we see in Exercise 4.7 that applying Lemma 4.2.1 to A has no hope of giving us this stronger conclusion.

The way around this is a clever idea introduced by Green and Ruzsa, who noted that if a large proportion of $A - A$ is contained in such a set then all of A is contained in such a set, and so applying Lemmas 4.2.1

and 4.2.2 to $A - A$ instead of to A gives us what we need. We capture Green and Ruzsa's idea in the following lemma.

Lemma 4.4.3 (Green–Ruzsa [35, Lemma 2.3]) *Let G be an abelian group, let $A \subset G$ be a finite set, and let $\varphi : G \to \mathbb{Z}/n\mathbb{Z}$ be a group homomorphism. Suppose that*

$$|(A - A) \setminus \varphi^{-1}[x, x + d]| < |A|/2 \tag{4.4.1}$$

for some $x \in \mathbb{Z}/n\mathbb{Z}$ and $d \in [0, n)$. Then $\varphi(A)$ is contained in an interval of width at most $2d + 1$.

Proof Upon translating A if necessary we may assume that $0 \in A$. This implies that $A \subset A - A$, and so (4.4.1) implies that $|A \setminus \varphi^{-1}[x, x + d]| < |A|/2$, and hence that $|A \cap \varphi^{-1}[x, x + d]| > |A|/2$. Picking an arbitrary $a \in A$, it therefore follows from (4.4.1) that the set $(A \cap \varphi^{-1}[x, x+d]) - a$ is not contained in $(A - A) \setminus \varphi^{-1}[x, x+d]$. Since $(A \cap \varphi^{-1}[x, x+d]) - a \subset A - A$, this means that there exists some $a' \in A \cap \varphi^{-1}[x, x+d]$ such that $\varphi(a' - a) \in [x, x + d]$, and hence $\varphi(a) \in \varphi(a') - [x, x + d] \subset [-d, d]$. □

Before we put all of this together to prove Proposition 4.1.5, we note the following rather crude estimate that will help with combining the various bounds we have so far derived.

Lemma 4.4.4 *Let $\delta \in (0, 1)$. Then for every $x \geq e^{\delta^{-2}}$ we have $\log x < x^{\delta}$.*

Proof We have $\log y < y^{1/2}$ for every $y > 0$, and so for $y \geq \delta^{-2}$ we have moreover that $\log y < y^{1/2} \leq \delta y$. Substituting $y = \log x$, this means that for $x \geq e^{\delta^{-2}}$ we have $\log(\log x) < \delta \log x = \log(x^{\delta})$, and the result follows. □

Proof of Proposition 4.1.5 By Lemma 4.4.1 we may assume that A is a subset of a finite abelian group G, and we may also assume that A has no s-model in any group smaller than G. To prove the proposition we must therefore show that $|G| \leq 4K^{20}(200s^2K^2)^{2K^{16}}|A|$.

Writing $\alpha = |A|/|G|$, suppose to the contrary that

$$\alpha < \frac{1}{4K^{20}(200s^2K^2)^{2K^{16}}}. \tag{4.4.2}$$

Theorem 2.3.1 implies that $|A - A|/|G| \leq K^2\alpha$ and $|2A - 2A| \leq K^4|A - A|$, and so Lemmas 4.2.1 and 4.2.2 imply that there exists a non-zero $\gamma \in \widehat{G}$ and an interval U of width $1/10s$ such that

$$|(A - A) \setminus \gamma^{-1}(U)| \leq 100s^2(2K^{10}\alpha)^{1/K^{16}} \log(\alpha^{-1}/K^2)|A - A|,$$

and hence, by Theorem 2.3.1, that

$$|(A-A)\setminus\gamma^{-1}(U)| \le 100K^2s^2(2K^{10}\alpha)^{1/K^{16}}\log(\alpha^{-1}/K^2)|A|. \quad (4.4.3)$$

Write $n = \operatorname{ord}(\gamma)$, and denote by C_n the subgroup of S^1 consisting of the integer powers of $e^{2\pi i/n}$. Let ψ_n be the unique homomorphism $C_n \to \mathbb{Z}/n\mathbb{Z}$ with $\psi_n(e^{2\pi i/n}) = 1$, and define a homomorphism $\varphi : G \to \mathbb{Z}/n\mathbb{Z}$ via $\varphi(g) = \psi_n(e(\gamma(g)))$. It then follows from (4.4.3) that there exists $x \in \mathbb{Z}/n\mathbb{Z}$ and $d \le n/10s$ such that

$$\begin{aligned}|(A-A)\setminus\varphi^{-1}[x,x+d]| &\le 100K^2s^2(2K^{10}\alpha)^{1/K^{16}}\log(\alpha^{-1}/K^2)|A| \\ &\le 100K^2s^2(2K^{10}\alpha)^{1/K^{16}}\log(\alpha^{-1})|A|. \quad (4.4.4)\end{aligned}$$

The bound (4.4.2) certainly implies that $\alpha \le e^{-K^{16}}$, and so Lemma 4.4.4 combines with (4.4.4) to imply that

$$|(A-A)\setminus\varphi^{-1}[x,x+d]| \le 100K^2s^2(2K^{10}\alpha^{1/2})^{1/K^{16}}|A|.$$

The assumption (4.4.2) then gives

$$|(A-A)\setminus\varphi^{-1}[x,x+d]| < |A|/2.$$

Lemma 4.4.3 therefore implies that $\varphi(A)$ is contained in an interval of width at most $2d+1 \le n/5s+1$, and then Lemma 4.4.2 implies that A has a Freiman s-model inside $(\ker\varphi)\times(\mathbb{Z}/(n-1)\mathbb{Z})$. This contradicts the assumption that A has no s-model inside a group smaller than G, and so contrary to (4.4.2) it must indeed have been the case that $|G| \le 4K^{20}(200s^2K^2)^{2K^{16}}|A|$, as required. □

4.5 Bohr Sets in Dense Subsets of Finite Abelian Groups

In this section we prove Proposition 4.1.4. The main ingredient is to show that $2A-2A$ contains a large Bohr set; as we saw in Chapter 3, it is then not too difficult to approximate this Bohr set by a coset progression, which is the structure we are ultimately aiming for. The following result is due to Green and Ruzsa, who built on ideas of Bogolyubov [8] in the case $G = \mathbb{Z}$.

Lemma 4.5.1 (Bogolyubov; Green–Ruzsa [35, §3]) *Let G be a finite abelian group, let $A \subset G$ be a subset of size $\alpha|G|$, and suppose that $|A+A| \le K|A|$. Let $\Gamma = \{\gamma \in \widehat{G} : |\widehat{1}_A(\gamma)| \ge \alpha/2K\}$. Then $|\Gamma| \le 4K^2/\alpha$ and $B(\Gamma, \frac{1}{6}) \subset 2A-2A$.*

The purpose of this section is to prove Lemma 4.5.1 and deduce Proposition 4.1.4 from it. In the next section we explain briefly how to improve Lemma 4.5.1 by enough to deduce Proposition 4.1.7, although we defer the details to Section 4.A.

In Exercise 4.2 we guide the reader to a simpler proof of Lemma 4.5.1 that does not require the small-doubling assumption, at the expense of worsening the dependence on α of the bounds.

Proof of Lemma 4.5.1 First, to bound $|\Gamma|$ note that $\sum_{\gamma\in\Gamma}|\widehat{1}_A(\gamma)|^2 \geq \alpha^2|\Gamma|/4K^2$ by definition of Γ. However, Parseval's identity (4.2.7) implies that

$$\sum_{\gamma\in\widehat{G}}|\widehat{1}_A(\gamma)|^2 = \mathbb{E}_{x\in G}|1_A(x)|^2 = \alpha,$$

and so we may conclude that $\alpha^2|\Gamma|/4K^2 \leq \alpha$, and hence that $|\Gamma| \leq 4K^2/\alpha$, as required.

The idea to prove that $B(\Gamma,\frac{1}{6}) \subset 2A-2A$ is to show that for every $x \in B(\Gamma,\frac{1}{6})$ we have $1_A * 1_{-A} * 1_A * 1_{-A}(x) > 0$; this implies that $x \in 2A-2A$ by (4.3.4). To see that it is true, first note that

$$\begin{aligned}
\sum_{\gamma\in\widehat{G}}|\widehat{1}_A(\gamma)|^4 &= \sum_{\gamma\in\widehat{G}}|\widehat{1_A * 1_{-A}}(\gamma)|^2 && \text{(by (4.3.3))}\\
&= \mathbb{E}_{x\in G}(1_A * 1_{-A}(x))^2 && \text{(by (4.2.7))}\\
&= \mathbb{E}_{x\in G}\left((1_A * 1_{-A}(x))^2 \cdot 1_{A-A}(x)\right) && \text{(by (4.3.4))}\\
&\geq \frac{(\mathbb{E}_{x\in G}\, 1_A * 1_{-A}(x))^2}{\mathbb{E}1_{A-A}} && \text{(by (1.5.1))}\\
&= \alpha^4/\mathbb{E}1_{A-A} && \text{(by (4.3.1))}\\
&\geq \alpha^3/K^2 && \text{(by Theorem 2.3.1)},
\end{aligned}$$

so that

$$\sum_{\gamma\in\widehat{G}}|\widehat{1}_A(\gamma)|^4 \geq \frac{\alpha^3}{K^2}. \tag{4.5.1}$$

Next, note from (4.3.2), (4.3.3) and (4.2.4) that for every $x \in G$ we have

$$1_A * 1_{-A} * 1_A * 1_{-A}(x) = \sum_{\gamma\in\widehat{G}}|\widehat{1}_A(\gamma)|^4 e(-\gamma(x)),$$

and hence, since $1_A * 1_{-A} * 1_A * 1_{-A}(x)$ is real,

$$1_A * 1_{-A} * 1_A * 1_{-A}(x) = \sum_{\gamma\in\widehat{G}}|\widehat{1}_A(\gamma)|^4\,\mathrm{Re}(e(\gamma(x))). \tag{4.5.2}$$

Now let $x \in B(\Gamma, \frac{1}{6})$, and note that $\mathrm{Re}(e(\gamma(x))) \geq 1/2$ for every $\gamma \in \Gamma$. Moreover, when $\gamma = 0$ we have $\mathrm{Re}(e(\gamma(x))) = 1 > 1/2$, and also $\widehat{1}_A(0) = \alpha$, and hence $0 \in \Gamma$ by definition of Γ. It therefore follows from (4.5.2) that

$$\begin{aligned}1_A * 1_{-A} * 1_A * 1_{-A}(x) &> \frac{1}{2}\sum_{\gamma\in\Gamma} |\widehat{1}_A(\gamma)|^4 - \sum_{\gamma\notin\Gamma} |\widehat{1}_A(\gamma)|^4 \\ &= \frac{1}{2}\sum_{\gamma\in\widehat{G}} |\widehat{1}_A(\gamma)|^4 - \frac{3}{2}\sum_{\gamma\notin\Gamma} |\widehat{1}_A(\gamma)|^4 \\ &\geq \frac{1}{2}\sum_{\gamma\in\widehat{G}} |\widehat{1}_A(\gamma)|^4 - \frac{3}{2}\sup_{\gamma\notin\Gamma} |\widehat{1}_A(\gamma)|^2 \sum_{\gamma\in\widehat{G}} |\widehat{1}_A(\gamma)|^2,\end{aligned}$$

and hence from (4.5.1) that

$$1_A * 1_{-A} * 1_A * 1_{-A}(x) > \frac{\alpha^3}{2K^2} - \frac{3}{2}\sup_{\gamma\notin\Gamma} |\widehat{1}_A(\gamma)|^2 \sum_{\gamma\in\widehat{G}} |\widehat{1}_A(\gamma)|^2.$$

Since $\sup_{\gamma\notin\Gamma} |\widehat{1}_A(\gamma)|^2 \leq \alpha^2/4K^2$ by definition of Γ and $\sum_{\gamma\in\widehat{G}} |\widehat{1}_A(\gamma)|^2 = \alpha$ by Parseval's identity (4.2.7), this implies that

$$\begin{aligned}1_A * 1_{-A} * 1_A * 1_{-A}(x) &> \frac{\alpha^3}{2K^2} - \frac{3\alpha^3}{8K^2} \\ &> 0,\end{aligned}$$

and hence by (4.3.4) that $x \in 2A - 2A$, as required. □

Proof of Proposition 4.1.4 This is immediate from Lemma 4.5.1 and Proposition 3.6.2. □

4.6 Reducing the Dimension of the Bohr Set

Chang [22] was able to reduce the rank of the progression given by Proposition 4.1.4 by reducing the dimension of the Bohr set given by Lemma 4.5.1. The basic idea of Chang's argument is to show that there is a small subset $\Gamma_0 \subset \Gamma$ that behaves a bit like a basis, in that every element of Γ can be expressed as a bounded linear combination of elements in Γ_0. This means that if $\gamma(x)$ is small for every $\gamma \in \Gamma_0$ then $\gamma(x)$ is also small for every $\gamma \in \Gamma$, and so there is some reasonably large ρ such that $B(\Gamma_0, \rho) \subset B(\Gamma, \frac{1}{6})$.

A basis in a finite-dimensional vector space can be constructed by taking a maximal set of linearly independent elements. In constructing our basis-like subset Γ_0 of Γ we use the following analogue of linear independence.

Definition 4.6.1 (dissociated subset of $\widehat{G}$) Let G be a finite abelian group, and let $\Gamma = \{\gamma_1, \ldots, \gamma_d\} \subset \widehat{G}$. The set Γ is said to be *dissociated* if the only solution to $\epsilon_1\gamma_1 + \cdots + \epsilon_d\gamma_d = 0$ with $\epsilon_i \in \{-1, 0, 1\}$ is the trivial solution with $\epsilon_i = 0$ for every i.

Just as a maximal set of linearly independent vectors in a finite-dimensional vector space spans that space, the following lemma exploits the fact that a maximal dissociated subset of a set $\Gamma \subset \widehat{G}$ spans Γ in some sense.

Lemma 4.6.2 *Let G be a finite abelian group, let $\Gamma \subset \widehat{G}$, and let $\Gamma_0 = \{\gamma_1, \ldots, \gamma_d\} \subset \Gamma$ be a maximal dissociated subset of Γ. Let $\rho \in [0, 1]$. Then we have*

$$B(\Gamma_0, \rho/d) \subset B(\Gamma, \rho).$$

Proof For an arbitrary $\gamma \in \Gamma \setminus \Gamma_0$ there exist, by the maximality of Γ_0, numbers $\epsilon_1, \ldots, \epsilon_d \in \{-1, 0, 1\}$ such that $\gamma = \epsilon_1\gamma_1 + \cdots + \epsilon_d\gamma_d$. For an arbitrary $x \in B(\Gamma_0, \rho/d)$, it follows that

$$\begin{aligned} |\gamma(x)| &\leq |\gamma_1(x)| + \cdots + |\gamma_d(x)| \\ &\leq \rho, \end{aligned}$$

and so $B(\Gamma_0, \rho/d) \subset B(\Gamma, \rho)$, as required. □

In proving Proposition 4.1.7, therefore, it is sufficient to bound the size of a maximal dissociated subset of the set Γ given by Lemma 4.5.1. One can think of this as eliminating some redundancy from Γ.

Proposition 4.6.3 (Chang [22]) *Let G be a finite abelian group, let $A \subset G$ be a subset of size $\alpha|G|$, and let $\rho \in [0, 1]$. Let $\Gamma = \{\gamma \in \widehat{G} : |\widehat{1}_A(\gamma)| \geq \rho\alpha\}$, and let $\Gamma_0 = \{\gamma_1, \ldots, \gamma_d\}$ be a dissociated subset of Γ. Then $d \leq 4\rho^{-2} \log \alpha^{-1}$.*

We prove Proposition 4.6.3 in the appendix to this chapter, Section 4.A.

Note in particular that Lemma 4.6.2 and Proposition 4.6.3 allow us to replace the Bohr set coming from Lemma 4.5.1, which has dimension linear in α^{-1}, with a Bohr set whose dimension is merely logarithmic in α^{-1}. This allows us to deduce Proposition 4.1.7, as follows.

Proof of Proposition 4.1.7 Lemma 4.5.1 implies that if $\Gamma = \{\gamma \in \widehat{G} : |\widehat{1}_A(\gamma)| \geq \alpha/2K\}$ then $|\Gamma| \leq 4K^2/\alpha$ and $B(\Gamma, \frac{1}{6}) \subset 2A - 2A$. Letting Γ_0 be a maximal dissociated subset of Γ, Proposition 4.6.3 implies that $|\Gamma_0| \leq 16K^2 \log \alpha^{-1}$, whilst Lemma 4.6.2 implies that

$$B(\Gamma_0, 1/96K^2 \log \alpha^{-1}) \subset B(\Gamma, \tfrac{1}{6}) \subset 2A - 2A.$$

The proposition is then immediate from Proposition 3.6.2. □

4.7 Chang's Covering Argument

In this section we prove Proposition 4.1.6, at the heart of which is the following covering argument due to Chang [22].

Proposition 4.7.1 (Chang's covering lemma) *Let G be a group. Suppose that $A \subset G$ is a finite subset of G satisfying $|A^{\epsilon_1} \cdots A^{\epsilon_n}| \leq K^n|A|$ for every $n \in \mathbb{N}$ and every $\epsilon_1, \ldots, \epsilon_n \in \{1, -1\}$. Let $m \in \mathbb{N}$, let*

$$\epsilon_1, \ldots, \epsilon_m \in \{1, -1\},$$

and suppose that $P \subset A^{\epsilon_1} \cdots A^{\epsilon_m}$ is a set satisfying $|P| \geq |A|/C$. Then there exist

$$t \leq \frac{(m+1)\log K + \log C}{\log 2} + 1 \tag{4.7.1}$$

and sets $S_1, \ldots, S_t \subset A$ with $|S_i| \leq 2K$ such that

$$A \subset S_t \cdots S_1 P P^{-1} S_1^{-1} \cdots S_{t-1}^{-1}. \tag{4.7.2}$$

Proof We define the sets S_i recursively. Assuming the sets $S_1, \ldots, S_i$ have already been defined, let $S_{i+1} \subset A$ be a maximal set of size at most $2K$ having the property that the translates $xS_i \cdots S_1P$ with $x \in S_{i+1}$ are disjoint. If $|S_{i+1}|$ is strictly less than $2K$ then stop and set $t = i+1$. The containment (4.7.2) then follows from Lemma 2.4.4, since S_t is maximal in A with respect to the property that the translates $xS_{t-1} \cdots S_1P$ with $x \in S_t$ are disjoint.

The disjointness of the various translates implies that

$$|S_t \cdots S_1 P| = |S_t| \cdots |S_1||P| \geq (2K)^{t-1}|P|,$$

while the fact that each $S_i \subset A$ and the assumption that $P \subset A^{\epsilon_1} \cdots A^{\epsilon_m}$ imply that

$$|S_t \cdots S_1 P| \leq K^{m+t}|A|.$$

Combining these two inequalities implies that $2^{t-1} \leq K^{m+1}|A|/|P| \leq CK^{m+1}$, which gives (4.7.1) and completes the proof. □

Proof of Proposition 4.1.6 Theorem 2.3.1 implies that $|mA - nA| \leq K^{m+n}|A|$ for every $m, n \geq 0$, and so Proposition 4.7.1 implies that there exist

$$t \leq \frac{(m+n+1)\log K + \log C}{\log 2} + 1$$

and sets $S_1, \ldots, S_t \subset A$ with $|S_i| \leq 2K$ such that

$$A \subset H + P - P \subset S_1 - S_1 + \cdots + S_{t-1} - S_{t-1} + S_t.$$

Enumerating the elements of $\bigcup_i S_i$ as $s_1, \ldots, s_r$ and writing

$$Q = \{\epsilon_1 s_1 + \cdots + \epsilon_r s_r : \epsilon_i, \in \{-1, 0, 1\}\},$$

we have $r \leq 2Kt$ and $A \subset P - P + Q \subset (m + n + 2Kt)(A - A)$, which completes the proof. □

4.A Dissociated Subsets of $\widehat{G}$

The aim of this appendix is to prove Proposition 4.6.3. We follow the approach taken by Green and Ruzsa [35, Proposition 3.2], starting with the following straightforward observation.

Lemma 4.A.1 *Let G, A, Γ and Γ_0 be as in Proposition 4.6.3. Then $d \leq \alpha^{-2}\rho^{-2}\sum_{i=1}^{d} |\widehat{1}_A(\gamma_i)|^2$.*

Proof The definition of Γ implies that $|\widehat{1}_A(\gamma_i)| \geq \rho\alpha$ for every i, from which the lemma is immediate. □

Bounding the sum $\sum_{i=1}^{d} |\widehat{1}_A(\gamma_i)|^2$ coming from Lemma 4.A.1 is slightly more involved. We start with the following lemma.

Lemma 4.A.2 *Let G, A, Γ and Γ_0 be as in Proposition 4.6.3, and define $f : G \to \mathbb{R}$ via*

$$f(x) = \sum_{i=1}^{d} c_i \operatorname{Re}(\widehat{1}_A(\gamma_i)e(-\gamma_i(x))),$$

with $c_i = 1$ if $\gamma_i \neq -\gamma_i$ and $c_i = \frac{1}{2}$ if $\gamma_i = -\gamma_i$. Then

$$\mathbb{E}_{x\in G} f(x)^2 = \frac{1}{2} \sum_{\{i:\gamma_i \neq -\gamma_i\}} |\widehat{1}_A(\gamma_i)|^2 + \frac{1}{4} \sum_{\{i:\gamma_i = -\gamma_i\}} |\widehat{1}_A(\gamma_i)|^2.$$

In particular, $\sum_{i=1}^d |\widehat{1}_A(\gamma_i)|^2 \le 4\mathbb{E}_{x\in G} f(x)^2$.

Proof We have

$$\begin{aligned}
\mathbb{E}_{x\in G} f(x)^2 &= \mathbb{E}_{x\in G} \sum_{i,j} c_i c_j \operatorname{Re}(\widehat{1}_A(\gamma_i)e(-\gamma_i(x))) \operatorname{Re}(\widehat{1}_A(\gamma_j)e(-\gamma_j(x))) \\
&= \frac{1}{4} \sum_{i,j} c_i c_j \, \mathbb{E}_{x\in G} \left(\widehat{1}_A(\gamma_i)e(-\gamma_i(x)) + \overline{\widehat{1}_A(\gamma_i)e(-\gamma_i(x))} \right) \\
&\qquad \times \left(\widehat{1}_A(\gamma_j)e(-\gamma_j(x)) + \overline{\widehat{1}_A(\gamma_j)e(-\gamma_j(x))} \right) \\
&= \frac{1}{4} \sum_{i,j} c_i c_j \, \mathbb{E}_{x\in G} \left(\widehat{1}_A(\gamma_i)e(-\gamma_i(x)) + \overline{\widehat{1}_A(\gamma_i)} e(\gamma_i(x)) \right) \\
&\qquad \times \left(\widehat{1}_A(\gamma_j)e(-\gamma_j(x)) + \overline{\widehat{1}_A(\gamma_j)} e(\gamma_j(x)) \right).
\end{aligned}$$

Whenever $i \neq j$ the fact that Γ_0 is dissociated implies that $\gamma_i \neq -\gamma_j$, and so the orthogonality relations (4.2.3) imply that

$$\mathbb{E}_{x\in G} e(\gamma_i(x) - \gamma_j(x)) = \mathbb{E}_{x\in G} e(\gamma_i(x) + \gamma_j(x)) = 0,$$

and we may conclude that

$$\begin{aligned}
&\mathbb{E}_{x\in G} f(x)^2 \\
&= \frac{1}{4} \sum_{i=1}^{d} c_i^2 \, \mathbb{E}_{x\in G} \left(\widehat{1}_A(\gamma_i)^2 e(-2\gamma_i(x)) + 2|\widehat{1}_A(\gamma_i)|^2 + \overline{\widehat{1}_A(\gamma_i)}^2 e(2\gamma_i(x)) \right).
\end{aligned} \tag{4.A.1}$$

If $\gamma_i = -\gamma_i$ then γ_i takes only the values 0 and $\frac{1}{2}$, and so all of the terms in the sum (4.2.5) defining the Fourier transform are real, and hence $\widehat{1}_A(\gamma_j) \in \mathbb{R}$. In particular, if $\gamma_i = -\gamma_i$ then

$$\widehat{1}_A(\gamma_i)^2 = \overline{\widehat{1}_A(\gamma_i)}^2 = |\widehat{1}_A(\gamma_i)|^2.$$

The lemma therefore follows from (4.A.1) and the orthogonality relations (4.2.3). □

We may rewrite the bound $4\mathbb{E}_{x\in G} f(x)^2$ coming from Lemma 4.A.2 as $4\sum_{\gamma\in\widehat{G}} |\widehat{f}(\gamma)|^2$ using Parseval's identity (4.2.7). In order to understand this bound we compute the Fourier transform of f, as follows.

Lemma 4.A.3 *Define $f : G \to \mathbb{R}$ as in Lemma 4.A.2. Let $\gamma \in \widehat{G}$. Then $\widehat{f}(\gamma) = \frac{1}{2}\widehat{1}_A(\gamma)$ if $\gamma \in \Gamma_0 \cup -\Gamma_0$, and $\widehat{f}(\gamma) = 0$ otherwise.*

Proof By (4.2.5) we have

$$\begin{aligned}
\widehat{f}(\gamma) &= \sum_{i=1}^{d} \mathbb{E}_{x\in G} e(\gamma(x)) c_i \operatorname{Re}(\widehat{1}_A(\gamma_i)e(-\gamma_i(x))) \\
&= \frac{1}{2}\sum_{i=1}^{d} c_i \mathbb{E}_{x\in G} e(\gamma(x)) \left(\widehat{1}_A(\gamma_i)e(-\gamma_i(x)) + \overline{\widehat{1}_A(\gamma_i)e(-\gamma_i(x))}\right) \\
&= \frac{1}{2}\sum_{i=1}^{d} c_i \mathbb{E}_{x\in G} e(\gamma(x)) \left(\widehat{1}_A(\gamma_i)e(-\gamma_i(x)) + \widehat{1}_A(-\gamma_i)e(\gamma_i(x))\right),
\end{aligned}$$

and so the lemma follows from (4.2.3), and from the fact that since Γ_0 is dissociated we have $\gamma_i \neq -\gamma_j$ when $i \neq j$. □

This in turn allows us to perform the following straightforward computation.

Lemma 4.A.4 *Define $f : G \to \mathbb{R}$ as in Lemma 4.A.2. Then*

$$\mathbb{E}_{x\in G} f(x)^2 = \tfrac{1}{2}\alpha\, \mathbb{E}_{x\in A} f(x).$$

Proof We have

$$\begin{aligned}
\mathbb{E}_{x\in G} f(x)^2 &= \sum_{\gamma\in\widehat{G}} |\widehat{f}(\gamma)|^2 && \text{(by (4.2.7))} \\
&= \tfrac{1}{2}\textstyle\sum_{\gamma\in\widehat{G}} \widehat{f}(\gamma)\overline{\widehat{1}_A(\gamma)} && \text{(by Lemma 4.A.3)} \\
&= \tfrac{1}{2}\mathbb{E}_{x\in G} f(x) 1_A(x) && \text{(by (4.2.6))} \\
&= \tfrac{1}{2}\alpha\, \mathbb{E}_{x\in A} f(x),
\end{aligned}$$

as required. □

Lemma 4.A.5 *Let $t \in \mathbb{R}$ and define $f : G \to \mathbb{R}$ as in Lemma 4.A.2. Then $\mathbb{E}_{x\in G} e^{tf(x)} \leq e^{t^2 \mathbb{E} f^2}$.*

Proof We have

$$\begin{aligned}
\mathbb{E}_{x\in G} e^{tf(x)} &= \mathbb{E}_{x\in G} e^{t\sum_{i=1}^{d} c_i \operatorname{Re}(\widehat{1}_A(\gamma_i)e(-\gamma_i(x)))} \\
&= \mathbb{E}_{x\in G} \prod_{i=1}^{d} e^{tc_i \operatorname{Re}(\widehat{1}_A(\gamma_i)e(-\gamma_i(x)))}.
\end{aligned}$$

Writing $\widehat{c}_i = c_i|\widehat{1}_A(\gamma_i)|$ and $\omega_i = \widehat{1}_A(\gamma_i)/|\widehat{1}_A(\gamma_i)|$ for each i, we may rewrite this as

$$\mathbb{E}_{x\in G}e^{tf(x)} = \mathbb{E}_{x\in G}\prod_{i=1}^{d} e^{t\widehat{c}_i \operatorname{Re}(\omega_i e(-\gamma_i(x)))}.$$

Since for every $y \in \mathbb{R}$ with $|y| \le 1$ we have $e^{ty} \le \cosh t + y \sinh t$, this in turn implies that

$$\begin{aligned}\mathbb{E}_{x\in G}e^{tf(x)} &\le \mathbb{E}_{x\in G}\prod_{i=1}^{d}(\cosh t\widehat{c}_i + \operatorname{Re}(\omega_i e(-\gamma_i(x)))\sinh t\widehat{c}_i)\\ &= \mathbb{E}_{x\in G}\prod_{i=1}^{d}\Big(\cosh t\widehat{c}_i + \tfrac{1}{2}(\omega_i e(-\gamma_i(x)) + \overline{\omega_i}e(+\gamma_i(x)))\sinh t\widehat{c}_i\Big).\end{aligned}$$

Multiplying out these brackets, we obtain a linear combination of terms of the form $e(\epsilon_1\gamma_1(x) + \cdots + \epsilon_d\gamma_d(x))$ with $\epsilon_i \in \{-1, 0, 1\}$ for each i. Since Γ_0 is dissociated, the orthogonality relations (4.2.3) imply that all these terms vanish when we take the expectation over $x \in G$ except the term with every $\epsilon_i = 0$, the coefficient of which is $\prod_{i=1}^{d}\cosh t\widehat{c}_i$, and so we conclude that

$$\mathbb{E}_{x\in G}e^{tf(x)} \le \prod_{i=1}^{d}\cosh t\widehat{c}_i.$$

Since $\cosh u \le e^{u^2/2}$, we conclude that

$$\begin{aligned}\mathbb{E}_{x\in G}e^{tf(x)} &\le \exp\left(\frac{1}{2}t^2\sum_{i=1}^{d}\widehat{c}_i^2\right)\\ &= \exp\left(t^2\left(\frac{1}{2}\sum_{\{i:\gamma_i\ne-\gamma_i\}}|\widehat{1}_A(\gamma_i)|^2 + \frac{1}{8}\sum_{\{i:\gamma_i=-\gamma_i\}}|\widehat{1}_A(\gamma_i)|^2\right)\right),\end{aligned}$$

which is at most $e^{t^2\mathbb{E}f^2}$ by Lemma 4.A.2, as required. □

Proof of Proposition 4.6.3 We use the arithmetic–geometric mean inequality,

$$\exp\left(\frac{x_1 + \cdots + x_n}{n}\right) \le \frac{e^{x_1} + \cdots + e^{e_n}}{n}. \tag{4.A.2}$$

Defining $f : G \to \mathbb{R}$ as in Lemma 4.A.2, for $t \in \mathbb{R}$ we have

$$\begin{aligned} e^{2t\mathbb{E}f^2/\alpha} &= e^{t\mathbb{E}_{x\in A}f(x)} && \text{(by Lemma 4.A.4)} \\ &\leq \mathbb{E}_{x\in A}e^{tf(x)} && \text{(by (4.A.2))} \\ &\leq \alpha^{-1}\mathbb{E}_{x\in G}e^{tf(x)} \\ &\leq \alpha^{-1}e^{t^2\mathbb{E}f^2} && \text{(by Lemma 4.A.5).} \end{aligned}$$

Taking $t = \alpha^{-1}$ and dividing through by $e^{\alpha^{-2}\mathbb{E}f^2}$, we conclude that

$$e^{\alpha^{-2}\mathbb{E}f^2} \leq \alpha^{-1}. \tag{4.A.3}$$

We then have

$$\begin{aligned} d &\leq \alpha^{-2}\rho^{-2}\sum_{i=1}^{d}|\widehat{1}_A(\gamma_i)|^2 && \text{(by Lemma 4.A.1)} \\ &\leq 4\alpha^{-2}\rho^{-2}\mathbb{E}_{x\in G}f(x)^2 && \text{(by Lemma 4.A.2)} \\ &\leq 4\rho^{-2}\log\alpha^{-1} && \text{(by (4.A.3)),} \end{aligned}$$

as required. □

Exercises

4.1 The purpose of this exercise is to show that in the case $A \subset \mathbb{Z}$, and at the expense of weakening the conclusion slightly (though not so much as to prevent it from being used in the proof of Theorem 4.1.3), one can prove Proposition 4.1.5 more simply and with better bounds. Throughout the exercise, given $m \in \mathbb{N}$ we write $\pi_m : \mathbb{Z} \to \mathbb{Z}/m\mathbb{Z}$ for the quotient homomorphism and $\psi_m : \mathbb{Z}/m\mathbb{Z} \to [m]$ for the unique map such that $\pi_m \circ \psi_m$ is the identity on $\mathbb{Z}/m\mathbb{Z}$. Given, in addition, $\lambda \in \mathbb{Z}/m\mathbb{Z}$, we write $d_{m,\lambda} : \mathbb{Z}/m\mathbb{Z} \to \mathbb{Z}/m\mathbb{Z}$ for the map defined by $x \mapsto \lambda x$.

(a) Let $m, s \in \mathbb{N}$, and suppose that $A \subset \mathbb{Z}/m\mathbb{Z}$ is such that $\psi_m(A)$ lies in an interval of length at most m/s. Show that ψ_m is a Freiman s-homomorphism on restriction to A.

(b) Let $A \subset \mathbb{N}$ be finite and let p be a prime greater than every element of $sA - sA$. Given $\lambda \in (\mathbb{Z}/p\mathbb{Z})^\times$, define $\varphi_\lambda : \mathbb{Z} \to \mathbb{Z}/m\mathbb{Z}$ via $\varphi_\lambda = \pi_m \circ \psi_p \circ d_{p,\lambda} \circ \pi_p$. Show that for each $\lambda \in (\mathbb{Z}/p\mathbb{Z})^\times$ there exists a subset $A' \subset A$ of size at least $|A|/s$ such that φ_λ is a Freiman s-homomorphism on restriction to A'.

(c) Given $\lambda \in (\mathbb{Z}/p\mathbb{Z})^\times$, show that if $\pi_m \circ \psi_p \circ \pi_p(\psi_p(\lambda)d) \neq 0$ for every $d \in sA - sA \setminus \{0\}$ then $\varphi_\lambda|_{A'} : A' \to \varphi_\lambda(A')$ is a Freiman s-isomorphism.

(d) Show that the number of $\lambda \in (\mathbb{Z}/p\mathbb{Z})^\times$ satisfying $\pi_m \circ \psi_p \circ \pi_p(\psi_p(\lambda)d) = 0$ for a given $d \in sA - sA \setminus \{0\}$ is at most $(p-1)/m$.

(e) Deduce that the number of $\lambda \in (\mathbb{Z}/p\mathbb{Z})^\times$ for which there exists $d \in sA - sA \setminus \{0\}$ such that $\pi_m \circ \psi_p \circ \pi_p(\psi_p(\lambda)d) = 0$ is at most

$$\frac{p-1}{m}(|sA - sA| - 1).$$

(f) Conclude that if $|A + A| \leq K|A|$ and $m = K^{16}|A|$ then there exists some $\lambda \in (\mathbb{Z}/p\mathbb{Z})^\times$ and some $A' \subset A$ of size at least $|A|/s$ such that $\varphi_\lambda(A')$ is a Freiman s-model for A' in $\mathbb{Z}/m\mathbb{Z}$.

4.2 Let G be a finite abelian group and let $A \subset G$ be an arbitrary subset of size $\alpha|G|$. Show that there exists a subset $\Gamma \subset \widehat{G}$ of size at most $4/\alpha^2$ such that $B(\Gamma, \frac{1}{6}) \subset 2A - 2A$. *This shows that a version of Lemma 4.5.1 can be proved more simply, and without using the small-doubling assumption, at the expense of worsening the dependence on α. Hint: First show that if $\Gamma = \{\gamma \in \widehat{G} : |\widehat{1}_A(\gamma)| \geq \frac{1}{2}\alpha^{3/2}\}$ then $|\Gamma| \leq 4/\alpha^2$. Then show that for $x \in B(\Gamma, \frac{1}{6})$ we have* $1_A * 1_{-A} * 1_A * 1_{-A}(x) > \frac{1}{2}|\widehat{1}_A(0)|^4 - \sup_{\gamma \notin \Gamma} |\widehat{1}_A(\gamma)|^2 \sum_{\gamma \in \widehat{G}} |\widehat{1}_A(\gamma)|^2$.

4.3 Prove Theorem 4.1.2 in the case $A \subset \mathbb{Z}$ using only Exercises 4.1 and 4.2 and material from earlier chapters. Understand why Exercise 4.2 cannot be used in place of Lemma 4.5.1 in the proof of Theorem 4.1.2 in general.

4.4 Sanders [59] has shown that in Theorem 4.1.3 the bound on the rank of P can be improved to $O(\log^{O(1)} 2K)$, and the bound on the cardinality of $H + P$ can be improved to

$$|H + P| \geq \exp(-O(\log^{O(1)} 2K))|A|.$$

Using these bounds, prove the following for a subset A of an abelian group satisfying $|A + A| \leq K|A|$.

(a) There exists $X \subset A$ of size at most $\exp(O(\log^{O(1)} 2K))$ and a coset progression $H + P \subset 4A - 4A$ of rank at most $O(\log^{O(1)} 2K)$ such that $A \subset X + H + P$.

(b) There exists a coset progression $H + P$ of rank at most $O(K \log^{O(1)} 2K)$ satisfying $A \subset H + P$ and

$$|H + P| \leq \exp(O(K \log^{O(1)} 2K))|A|.$$

4.5 Let $n \in \mathbb{N}$, and suppose that $A \subset \mathbb{F}_2^n$ satisfies $|A+A| \leq K|A|$. Show that A has a Freiman 2-model in a group of size at most $K^4|A|$. *This improves Proposition 4.1.5 in the case $s = 2$ for subsets of $\mathbb{F}_2^n$. Hint: If $|\mathbb{F}_2^n| > K^4|A|$ then Theorem 2.3.1 implies that there exists $x \in \mathbb{F}_2^n \setminus (2A - 2A)$. Show that the linear projection $\varphi : \mathbb{F}_2^n \to \mathbb{F}_2^n/\{0, x\}$ is a Freiman 2-isomomorphism on restriction to A.*

4.6 Let G be a finite abelian group and let $A \subset G$. Show that if A is contained in a coset of a proper subgroup of G then there is some non-zero $\gamma \in \widehat{G}$ such that

$$|\widehat{1}_A(\gamma)| = \frac{|A|}{|G|}.$$

Conversely, given $\rho \in (0, 1)$ show that if there exists a non-zero $\gamma \in \widehat{G}$ such that

$$|\widehat{1}_A(\gamma)| \geq \left(1 - \frac{\rho}{|G|^2}\right)\frac{|A|}{|G|}$$

then a proportion of at least $(1 - \rho)$ of the elements of A belong to a single coset of some proper subgroup of G.

4.7 Let $\eta \in (0, 1)$. Show that if $m \in \mathbb{N}$ is large enough depending on η then there exist $A \subset \mathbb{Z}/m\mathbb{Z}$ and $\gamma \in \widehat{\mathbb{Z}/m\mathbb{Z}}$ such that

$$|\widehat{1}_A(\gamma)| \geq (1 - \eta)\frac{|A|}{|G|},$$

but such that A is not contained in $\gamma^{-1}(U)$ for any interval $U \subset \mathbb{T}$ of width less than $\frac{1}{2}$. *Thus, the conclusion of Lemma 4.2.2 cannot be strengthened to say that $A \subset \gamma^{-1}(U)$. Hint: Let A be the union of an interval and a singleton.*

4.8 Show that if $d < n/3$ in Lemma 4.4.3 then the conclusion can be strengthened to say that A lies in an interval of length at most $d + 1$.

5

Nilpotent Groups, Commutators and Nilprogressions

5.1 Progressions in the Heisenberg Group

The Freiman–Green–Ruzsa theorem of the last chapter gave us a qualitatively complete description of the sets of small doubling in an arbitrary abelian group. We now turn our attention to sets of small doubling in non-abelian groups.

We start in a particular non-abelian group called the *Heisenberg group*. The Heisenberg group $H(\mathbb{Z})$ is a group of 3×3 matrices defined via

$$H(\mathbb{Z}) = \begin{pmatrix} 1 & \mathbb{Z} & \mathbb{Z} \\ 0 & 1 & \mathbb{Z} \\ 0 & 0 & 1 \end{pmatrix} = \left\{ \begin{pmatrix} 1 & n_2 & n_3 \\ 0 & 1 & n_1 \\ 0 & 0 & 1 \end{pmatrix} : n_i \in \mathbb{Z} \right\}.$$

It is easy to see that if we write

$$u_1 = \begin{pmatrix} 1 & 0 & 0 \\ 0 & 1 & 1 \\ 0 & 0 & 1 \end{pmatrix}, u_2 = \begin{pmatrix} 1 & 1 & 0 \\ 0 & 1 & 0 \\ 0 & 0 & 1 \end{pmatrix}, u_3 = \begin{pmatrix} 1 & 0 & 1 \\ 0 & 1 & 0 \\ 0 & 0 & 1 \end{pmatrix} \tag{5.1.1}$$

then every element of $H(\mathbb{Z})$ can be expressed in the form

$$\begin{pmatrix} 1 & n_2 & n_3 \\ 0 & 1 & n_1 \\ 0 & 0 & 1 \end{pmatrix} = u_1^{n_1} u_2^{n_1} u_3^{n_3}, \tag{5.1.2}$$

and, given two elements $u_1^{n_1} u_2^{n_2} u_3^{n_3}$ and $u_1^{n_1'} u_2^{n_2'} u_3^{n_3'}$ of $H(\mathbb{Z})$ expressed in this way we have an explicit formula for their product:

$$(u_1^{n_1} u_2^{n_2} u_3^{n_3})(u_1^{n_1'} u_2^{n_2'} u_3^{n_3'}) = u_1^{n_1+n_1'} u_2^{n_2+n_2'} u_3^{n_3+n_3'+n_1' n_2}. \tag{5.1.3}$$

The formula (5.1.3) is trivial to verify using the matrix representation of $H(\mathbb{Z})$, but there is also a slightly more abstract explanation for it.

We define the *commutator* $[x, y]$ of two elements x, y in some group via $[x, y] = x^{-1}y^{-1}xy$. In light of the trivial identity

$$yx = xy[y, x], \tag{5.1.4}$$

we think of the commutator $[y, x]$ as representing the 'cost' of interchanging the order of x and y in a word in the elements of G. In an abelian group, the fact that commutators are trivial corresponds to the fact that elements in a word can be rearranged freely without affecting the value of the word.

In the Heisenberg group one may check that $[u_2, u_1] = u_3$, so given a word in u_1 and u_2 we can, at the cost of introducing a copy of $[u_2, u_1] = u_3$ to the word, interchange the order of an adjacent pair of letters u_1 and u_2. Rewriting the formula (5.1.3) as

$$(u_1^{n_1}u_2^{n_2}[u_2, u_1]^{n_3})(u_1^{n_1'}u_2^{n_2'}[u_2, u_1]^{n_3'}) = u_1^{n_1+n_1'}u_2^{n_2+n_2'}[u_2, u_1]^{n_3+n_3'+n_1'n_2}, \tag{5.1.5}$$

it becomes clear that the $n_1'n_2$ term in the exponent of $[u_2, u_1]$ corresponds to the fact that in rearranging the elements on the left-hand side to obtain the right-hand side one must interchange the positions of n_1' copies of u_1 with n_2 copies of u_2.

We will now look to generalise to the Heisenberg group the abelian notion of progression we introduced in Definition 3.1.1. One naive attempt at generalising that notion to non-abelian groups is the following.

Definition 5.1.1 (ordered progression) Let $x_1, \dots, x_r$ be elements in a group G, and let $L_1, \dots, L_r \geq 0$. Then the *ordered progression* $P_{\text{ord}}(x; L) = P_{\text{ord}}(x_1, \dots, x_r; L_1, \dots, L_r)$ is defined to be

$$P_{\text{ord}}(x; L) = \{x_1^{\ell_1} \cdots x_r^{\ell_r} : |\ell_i| \leq L_i\}.$$

We define r to be the *rank* of $P_{\text{ord}}(x; L)$.

However, it is fairly easy to see that ordered progressions do not in general have small doubling in non-abelian groups. Indeed, defining u_1 and u_2 as in (5.1.1) and letting $L_1, L_2 \geq 0$, we have

$$P_{\text{ord}}(u; L) = \{u_1^{\ell_1}u_2^{\ell_2} : |\ell_i| \leq L_i\}.$$

Since

$$(u_1^{\ell_1}u_2^{\ell_2})(u_1^{\ell_1'}u_2^{\ell_2'}) = u_1^{\ell_1+\ell_1'}u_2^{\ell_2+\ell_2'}u_3^{\ell_1'\ell_2}, \tag{5.1.6}$$

and since by changing the values ℓ_1, ℓ_1' within the given ranges one can obtain many different values of the entry $\ell_1'\ell_2$, it is not difficult to see that $|P_{\text{ord}}(u; L)^2|/|P_{\text{ord}}(u; L)| \to \infty$ as $L_1, L_2 \to \infty$.

One can think of this as being an extra 'degree of freedom' in the progression $P_{\mathrm{ord}}(u;L)^2$ compared to $P_{\mathrm{ord}}(u;L)$, represented by the power of u_3. This extra degree of freedom essentially comes from the existence of a certain amount of freedom in the order in which the $\ell_1+\ell_1'$ instances of u_1 and the $\ell_2+\ell_2'$ instances of u_2 appear on the left-hand side of (5.1.6). By contrast, the definition of ordered progression gives no freedom in the order in which the instances of u_1 and u_2 appear in an element of $P_{\mathrm{ord}}(u;L)$ itself.

It turns out that by introducing to the original progression the same freedom to reorder the elements, one can force it to have small doubling after all.

Definition 5.1.2 (non-abelian progression) Let $x_1,\dots,x_r$ be elements in a group G, and let $L_1,\dots,L_r \geq 0$. Then the *non-abelian progression* $P(x;L) = P(x_1,\dots,x_r;L_1,\dots,L_r)$ is defined to consist of those elements of G that are expressible as products of the elements $x_1^{\pm 1},\dots,x_r^{\pm 1}$ in which each element x_i and its inverse appear at most L_i times between them. We define r to be the *rank* of $P_{\mathrm{ord}}(x;L)$.

Proposition 5.1.3 *Let u_1,u_2 be as in* (5.1.1) *and let $L_1,L_2 \geq 0$. Then* $|P(u;L)^3| \ll |P(u;L)|$.

Lemma 5.1.4 *Let u_1,u_2 be as in* (5.1.1), *and define*

$$\overline{P}(u;L) = \{u_1^{\ell_1}u_2^{\ell_2}[u_2,u_1]^{\ell_3} : |\ell_1| \leq L_1, |\ell_2| \leq L_2, |\ell_3| \leq L_1L_2\}.$$

Then $P(u;L) \subset \overline{P}(u;L) \subset P(u;5L)$.

Proof One can easily verify that

$$[u_2^{n_2},u_1^{n_1}] = [u_2,u_1]^{n_1n_2} \tag{5.1.7}$$

for every $n_1,n_2 \in \mathbb{Z}$. The first inclusion of the lemma then follows from repeated use of identity (5.1.4), the cases of (5.1.7) in which $n_i = \pm 1$, and the fact that $[u_2,u_1]$ is central in $H(\mathbb{Z})$.

To prove the second inclusion of the lemma, given $\ell_1,\ell_2,\ell_3 \in \mathbb{Z}$ satisfying

$$\begin{aligned} |\ell_1| &\leq L_1, \\ |\ell_2| &\leq L_2, \\ |\ell_3| &\leq L_1L_2, \end{aligned}$$

write $\ell_3 = mL_2+r$ with $-L_1 \leq m \leq L_1$ and $0 \leq r \leq L_2$, and use (5.1.7) again to conclude that

$$u_1^{\ell_1} u_2^{\ell_2} [u_2, u_1]^{\ell_3} = u_1^{\ell_1} u_2^{\ell_2} [u_2^{L_2}, u_1^m][x_2^r, x_1],$$

which clearly lies in $P(u; 5L)$. □

Proof of Proposition 5.1.3 The proposition is trivial if $L_1 < 1$ or $L_2 < 1$, so we may assume that $L_1, L_2 \geq 1$. Lemma 5.1.4 implies in particular that $\overline{P}(u; \frac{1}{5}L) \subset P(u; L)$ and

$$P(u; L)^3 \subset P(u; 3L) \subset \overline{P}(u; 3L) \tag{5.1.8}$$

for every $L_1, L_1 \in \mathbb{N}$.

We first prove the proposition in the case $L_1, L_2 \geq 5$, in which it follows from the above containments that

$$\overline{P}(x; \lfloor \tfrac{1}{5}L \rfloor) \subset P(u; L)$$

and

$$P(u; L)^3 \subset \overline{P}(u; 30\lfloor \tfrac{1}{5}L \rfloor).$$

The proposition then follows from the fact that

$$|\overline{P}(u; 30L')| \leq 30^4 |\overline{P}(u; 30L')|$$

for every $L'_1, L'_2 \in \mathbb{N}$.

If either $L_1 < 5$ or $L_2 < 5$ then we may verify the proposition directly. We will treat the case in which $L_2 < 5$; the case in which $L_1 < 5$ is very similar. Suppose, then, that $L_2 < 5$. Note that since $L_2 \geq 1$, each of the elements $u_1^\ell u_2 u_1^m$ with $|\ell| + |m| \leq L_1$ belongs to $P(u; L)$. Since $u_1^\ell u_2 u_1^m = u_1^{\ell+m} u_2 u_3^m$, these elements are all distinct by (5.1.2), and so $|P(u; L)| \gg L_1^2$. However, since $L_2 < 5$ we also have $|\overline{P}(u; 3L)| \ll L_1^2$, and so the proposition follows from (5.1.8). □

At this point it is worth cautioning that even non-abelian progressions are not necessarily sets of small doubling in an arbitrary group: consider, for example, the progression $P(x_1, x_2; L_1, L_2)$ when x_1, x_2 are generators of a free group. The property of $H(\mathbb{Z})$ that allowed us to prove Proposition 5.1.3 was that commutators are central in $H(\mathbb{Z})$. This property is in fact a specific instance of a more general property called *nilpotence*. We spend the rest of this chapter introducing nilpotent groups in general, and showing that non-abelian progressions in nilpotent groups – which we will call *nilprogressions* – are always sets of small tripling.

5.2 Nilpotent Groups

In this section we introduce general nilpotent groups. Some of the material is reproduced from Hall [38, Chapter 10].

Definition 5.2.1 (normal and central series) Given a group G, we define a (possibly finite) series $G = H_1 > H_2 > \cdots$ of subgroups of G to be a *normal series* for G if each subgroup H_i is normal in G. We define a normal series $G = H_1 > H_2 > \cdots$ to be a *central series* for G if each H_i/H_{i+1} is central in the quotient G/H_{i+1}.

Definition 5.2.2 (nilpotent group) A group G is *nilpotent* if it admits a finite central series $G = H_1 > \cdots > H_{s+1} = \{1\}$. We define the *step* (also called the *class*) of a nilpotent group G to be the smallest s for which such a series exists.

Thus, for example, a group is 1-step nilpotent if and only if it is abelian. It is easy to verify that in the Heisenberg group $H(\mathbb{Z})$ commutators are central, and hence that the Heisenberg group is 2-step nilpotent.

Given elements $x_1, \ldots, x_k$ of a group G we define the *simple commutator* $[x_1, \ldots, x_k]_k$ recursively via $[x_1]_1 = x_1$ and

$$[x_1, \ldots, x_k]_k = [[x_1, \ldots, x_{k-1}]_{k-1}, x_k] \tag{5.2.1}$$

for $k \geq 2$. We define the commutator $[H_1, H_2]$ of two subgroups H_i of a group G via

$$[H_1, H_2] = \langle [h_1, h_2] : h_i \in H_i \rangle,$$

and then similarly, given subgroups $H_1, \ldots, H_k$ of a group G we define the *simple commutator* $[H_1, \ldots, H_k]_k$ recursively via $[H_1]_1 = H_1$ and

$$[H_1, \ldots, H_k]_k = [[H_1, \ldots, H_{k-1}]_{k-1}, H_k]$$

for $k \geq 2$. We drop the subscript k from the brackets when it is clear from the context what value it takes, such as is the case for $[x_1, \ldots, x_k] = [x_1, \ldots, x_k]_k$, for example (note, though, that dropping the subscript k from the commutator $[G, \ldots, G]_k$ would create ambiguity).

Note that for two subgroups $H_1, H_2 < G$ we have

$$[H_1, H_2] = [H_2, H_1], \tag{5.2.2}$$

since $[x, y] = [y, x]^{-1}$ for arbitrary $x, y \in G$.

Lemma 5.2.3 *Let G be a group, let $N, H_1, \dots, H_k \lhd G$ be normal subgroups of G, and for each i let S_i be a generating set for H_i. Suppose that $[s_1, \dots, s_k] \in N$ whenever $s_i \in S_i$. Then $[H_1, \dots, H_k] \subset N$.*

Proof We proceed by induction on k, the case $k = 1$ being trivial. Suppose, therefore, that $k > 1$ and that $[s_1, \dots, s_k] \in N$ whenever $s_i \in S_i$. This implies in particular that $[s_1, \dots, s_{k-1}] \in C_{G/N}(H_k)$ whenever $s_i \in S_i$. Since H_k is normal, $C_{G/N}(H_k)$ is also normal by Lemma 1.5.3, so we conclude by induction that $[H_1, \dots, H_{k-1}] \subset C_{G/N}(H_k)$ and the lemma is proved. □

We define the *lower central series* $G = G_1 > G_2 > \cdots$ of an arbitrary group G by setting

$$G_k = \langle [x_1, \dots, x_k] : x_i \in G \rangle$$

for each k. The fact that $G_k > G_{k+1}$ follows easily from the fact that $[x_1, \dots, x_k] = [[x_1, x_2] \dots, x_k]$. Note that the subgroups G_k are not only normal, but characteristic in G. The lower central series is thus certainly a normal series. As the name suggests it is also a central series; this is not quite so obvious, but is an immediate consequence of the following result.

Proposition 5.2.4 *If $G = G_1 > G_2 > \cdots$ is the lower central series of a group G then $G_{k+1} = [G_k, G]$ for every k. In particular, $[G, \dots, G]_k = G_k$ for every k.*

Proof The fact that $G_{k+1} < [G_k, G]$ is trivial. On the other hand, since the groups G_i are normal the fact that $[G_k, G] < G_{k+1}$ follows from Lemma 5.2.3. □

Corollary 5.2.5 *The lower central series of an arbitrary group is a central series.*

The following proposition says that if S is a generating set for G then G_k is generated modulo G_{k+1} by the simple commutators in the elements of S.

Proposition 5.2.6 *Let G be a group with generating set S. Then*

$$G_k = \langle [s_1, \dots, s_k] G_{k+1} : s_i \in S \rangle.$$

Proof We have $\langle [s_1, \dots, s_k] G_{k+1} : s_i \in S \rangle \subset G_k$ by definition. On the other hand, given $g \in G$ and $s_1, \dots, s_k \in S$, since $[[s_1, \dots, s_k], g] \in G_{k+1}$ by definition, we have $[s_1, \dots, s_k]^g \in [s_1, \dots, s_k] G_{k+1}$. Since G_{k+1} is normal in G, it follows that $\langle [s_1, \ldots, s_k]G_{k+1} : s_i \in S \rangle$ is also normal in G. We therefore have

$$\begin{aligned} G_k &= [G, \ldots, G]_k && \text{(by Lemma 5.2.4)} \\ &\subset \langle [s_1, \ldots, s_k]G_{k+1} : s_i \in S \rangle && \text{(by Lemma 5.2.3),} \end{aligned}$$

and so the proposition is proved. □

For the next proposition we need the commutator identity

$$[x, y^{-1}, z]^y [y, z^{-1}, x]^z [z, x^{-1}, y]^x = 1, \tag{5.2.3}$$

which may be verified easily by direct computation.

Proposition 5.2.7 *Let G be a group, and let $G = G_1 > G_2 > \cdots$ be the lower central series of G. Then $[G_i, G_j] \subset G_{i+j}$ for every $i, j \in \mathbb{N}$.*

Proof The case $j = 1$ is given by Proposition 5.2.4, so we may assume that $j > 1$ and, by induction, that

$$[G_k, G_{j-1}] \subset G_{k+j-1} \tag{5.2.4}$$

for every $k \in \mathbb{N}$. Note that

$$[G_i, G_j] = [G_i, [G_{j-1}, G]] = [[G, G_{j-1}], G_i] \tag{5.2.5}$$

by Proposition 5.2.4 and (5.2.2). Note, moreover, that

$$[[G_i, G], G_{j-1}] = [G_{i+1}, G_{j-1}] \subset G_{i+j} \tag{5.2.6}$$

by Proposition 5.2.4 and (5.2.4), and

$$[[G_{j-1}, G_i], G] \subset [G_{i+j-1}, G] \subset G_{i+j} \tag{5.2.7}$$

by (5.2.2), (5.2.4) and Proposition 5.2.4.

Given $x \in G$, $y \in G_{j-1}$ and $z \in G_i$, we have

$$\begin{aligned} [x, y, z] &= (([y^{-1}, z^{-1}, x]^z [z, x^{-1}, y^{-1}]^x)^{-1})^y && \text{(by (5.2.3))} \\ &\in G_{i+j} && \text{(by (5.2.6) and (5.2.7)).} \end{aligned}$$

Lemma 5.2.3 therefore implies that $[[G, G_{j-1}], G_i] \subset G_{i+j}$, and so the proposition follows from (5.2.5). □

We next define the *upper central series* $\{1\} = Z_0(G) < Z_1(G) < Z_2(G) < \cdots$ of an arbitrary group G by setting each $Z_{i+1}(G)$ so that $Z_{i+1}(G)/Z_i(G)$ is the centre of $G/Z_i(G)$. Since the centre of a group is characteristic, each of the groups $Z_i(G)$ is characteristic in G. Note that the upper central series is not a central series in the sense of Definition 5.2.1 unless it is finite, in which case G is nilpotent.

$$
\begin{array}{ccccccccccc}
 & & Z_r(G) & > & Z_{r-1}(G) & > & \cdots & > & Z_1(G) & > & Z_0(G) = \{1\} \\
 & & \vee & & \vee & & & & \vee & & \vee \\
 & & H_1 & > & H_2 & > & \cdots & > & H_r & > & H_{r+1} \\
 & & \vee & & \vee & & & & \vee & & \vee \\
G & = & G_1 & > & G_2 & > & \cdots & > & G_r & > & G_{r+1}
\end{array}
$$

Figure 5.1 Illustration of Proposition 5.2.8 and the terms *upper* and *lower* central series

The prefixes *lower* and *upper* in the terms lower and upper central series are justified by the following result, which is illustrated in Figure 5.1.

Proposition 5.2.8 *Let G be a nilpotent group, and suppose that $G = H_1 > \cdots > H_{r+1} = \{1\}$ is a central series for G. Then $H_i \supset G_i$ for every $i = 1, \ldots, r+1$ and $H_{r+1-j} \subset Z_j(G)$ for every $j = 0, \ldots, r$.*

Corollary 5.2.9 *If a group G is nilpotent of step exactly s then both its upper and lower central series have length exactly $s+1$, in the sense that $G_s \neq G_{s+1} = \{1\}$ and $Z_{s-1}(G) \neq Z_s(G) = G$.*

Proof of Proposition 5.2.8 We first prove that for $i = 1, \ldots, r+1$ we have $H_i \supset G_i$, the case $i = 1$ being true by definition. For $i > 1$, we may assume by induction that

$$H_{i-1} \supset G_{i-1}. \tag{5.2.8}$$

The fact that H_{i-1}/H_i is central in G/H_i then means that

$$
\begin{aligned}
H_i &\supset [H_{i-1}, G] & & \\
&\supset [G_{i-1}, G] & & \text{(by (5.2.8))} \\
&= G_i & & \text{(by Proposition 5.2.4)},
\end{aligned}
$$

as required.

We now prove that $H_{r+1-j} \subset Z_j(G)$ for every $j = 0, \ldots, r$, the case $j = 0$ being true by definition. For $j > 0$, we may assume by induction that $H_{r+2-j} \subset Z_{j-1}(G)$, and hence that

$$G/Z_{j-1}(G) = \frac{G/H_{r+2-j}}{Z_{j-1}(G)/H_{r+2-j}}. \tag{5.2.9}$$

Now H_{r+1-j}/H_{r+2-j} is central in G/H_{r+2-j} by definition, so its image in $G/Z_{j-1}(G)$ in the quotient (5.2.9) is also central. Since $Z_j(G)/Z_{j-1}(G)$ is the centre of $G/Z_{j-1}(G)$ by definition, this implies that $H_{r+1-j} \subset Z_j(G)$, as required. □

The key points for the reader to take away from this section are as follows:

- A group G is nilpotent of step at most if and only if $G_{s+1} = \{1\}$, and if and only if $Z_s(G) = G$.
- In order to check that a group G with a generating set S is nilpotent of step at most s, it suffices to check that $[x_1, \ldots, x_{s+1}] = 1$ whenever $x_i \in S$.
- In the event that G is nilpotent of step at most s, commutators of commutators such as $[[[g_1, g_2], g_3], [g_4, g_5]]$ always evaluate to the identity if they include more than s elements $g_i \in G$ (in the language of Section 5.3, every commutator of *total weight* greater than s vanishes; see Lemma 5.3.2 for a formal statement).

5.3 Commutators

The fact that commutators are central in the Heisenberg group makes the formula (5.1.5) considerably simpler, since it means that when we interchange the order of x_2 and x_1 at the cost of introducing a copy of $[x_2, x_1]$, we are free to shift that new copy of $[x_2, x_1]$ to the right in order to end up with an expression of the form

$$x_1^{n_1} x_2^{n_2} [x_2, x_1]^{n_3}. \tag{5.3.1}$$

In more general groups than $H(\mathbb{Z})$, however, interchanging the order of a new instance of a commutator $[x_2, x_1]$ and an element x_1, say, results in an error of the form $[[x_2, x_1], x_1]$, and interchanging this commutator with an element x_1 results in an error of the form $[[[x_2, x_1], x_1], x_1]$, and so on.

The aims of this section are to give a formal definition of these iterated commutators, and to present some of their basic properties. In the next section we shall use this to derive a suitable analogue of the expression (5.3.1) valid in an arbitrary nilpotent group.

Definition 5.3.1 (commutators and weights) Let $x_1, \ldots, x_r$ be a finite set of symbols, which we shall call *letters*. We recursively define the *commutators* in the letters $x_1, \ldots, x_r$ by defining each x_i to be a commutator, and for every pair α, α' of commutators defining $[\alpha, \alpha']$ also to be a commutator. To each commutator α we assign a *weight vector* $\chi(\alpha) = (\chi_1(\alpha), \ldots, \chi_r(\alpha))$, defined recursively by setting

$$\chi_i(x_j) = \begin{cases} 1 & \text{if } i = j \\ 0 & \text{if } i \neq j \end{cases}$$

and, given two commutators α, α' in the x_j, defining

$$\chi([\alpha, \alpha']) = \chi(\alpha) + \chi(\alpha').$$

We call $\chi_i(\alpha)$ the *weight* of x_i in α, or the *x_i-weight* of α. We define the *total weight* $|\chi(\alpha)|$ of a commutator α via

$$|\chi(\alpha)| = \chi_1(\alpha) + \cdots + \chi_r(\alpha).$$

We define the *degree* $\deg(\alpha)$ of a commutator α via

$$\deg(\alpha) = \max_i \chi_i(\alpha),$$

and we call a degree-1 commutator a *linear* commutator. We define a commutator $[\alpha, \alpha']$ to be a *trivial commutator* if $\alpha = \alpha'$ or if either α or α' is trivial.

Thus, for example, $[[[x_2, x_1], x_1]]$ is a commutator in x_1, x_2 with weight vector $(2, 1)$, total weight 3 and degree 2.

Of course, if the letters x_i are elements that generate a group G then we may interpret commutators recursively via

$$[\alpha, \beta] = \alpha^{-1}\beta^{-1}\alpha\beta. \tag{5.3.2}$$

It is easy to see that a trivial commutator always has the identity element as its interpretation. If G is s-step nilpotent then those commutators of total weight greater than s also have trivial interpretations in G, as follows.

Lemma 5.3.2 *In an s-step nilpotent group every commutator of total weight greater than s evaluates to the identity.*

Proof This is immediate from Proposition 5.2.7 and Corollary 5.2.9. □

Proposition 5.3.3 *The number of commutators of weight at most s in the letters $x_1, \ldots, x_r$ is at most $(4r)^s$.*

Before we prove Proposition 5.3.3 we isolate two lemmas.

Lemma 5.3.4 *A commutator α of total weight k uses $k-1$ pairs of brackets.*

Proof The lemma is trivial for commutators of total weight 1. We may suppose, therefore, that $\alpha = [\beta_1, \beta_2]$ for some commutators β_1 and β_2 of total weights k_1 and k_2, respectively, with $k_1 + k_2 = k$. By induction, the number of pairs of brackets used by α is therefore $(k_1-1)+(k_2-1)+1 = k-1$, as required. □

Lemma 5.3.5 *Let $r, k \in \mathbb{N}$. Write $C_{r,k}$ for the set of commutators of weight k in the letters $x_1, \ldots, x_r$. Write W_r for the set of finite strings of the open-bracket symbol $[$ and the symbols $x_1, \ldots, x_r$. Write $W_{r,k}$ for the subset of W consisting of those strings that feature exactly $k-1$ copies of the symbol $[$ and exactly k copies of the symbols $x_1, \ldots, x_r$ between them. Then $|C_{r,k}| \leq |W_{r,k}|$.*

Proof We will define an injection $\psi : C_{r,k} \to W_{r,k}$. We may view a commutator $c \in C$ as a string in the symbols $x_1, \ldots, x_r$, the open- and close-bracket symbols $[$ and $]$, and the comma symbol. We may then define $\psi(c)$ to be the string obtained from c by deleting from c all copies of the symbol $]$ and all commas. Thus, for example, $\psi([[x_1, x_2], x_1]) = [[x_1 x_2 x_1$. The fact that $\psi(c) \in W_{r,k}$ follows from Lemma 5.3.4.

The case $k = 1$ is trivial, since $C_{r,1} = W_{r,1} = \{x_1, \ldots, x_r\}$ and ψ is the identity in that case. We may therefore assume that $k \geq 2$ and, by induction, that the lemma holds for all values of r for all smaller values of k.

We will show that knowledge of $\psi(c)$ allows us to recover c. Since $k \geq 2$, the string $c \in C_{r,k}$ must feature at least one substring of the form $[x_i, x_j]$, and hence $\psi(c)$ must feature at least one substring of the form $[x_i x_j$. Note also that every substring of $\psi(c)$ of the form $[x_i x_j$ must have arisen from a substring of c of the form $[x_i, x_j]$. Replacing the left-most substring of the form $[x_i, x_j]$ in c by a new symbol x_{r+1}, we obtain an element $c' \in C_{r+1,k-1}$. Note that we can obtain $\psi(c')$ directly from $\psi(c)$ by replacing the left-most substring of the form $[x_i x_j$ in $\psi(c)$ by x_{r+1}.

By induction on k, we can recover c' from $\psi(c)$. Since knowledge of $\psi(c)$ allows us to recover the substring $[x_i, x_j]$ of c that was replaced by x_{r+1} to arrive at c', we can then also recover c. This proves the lemma. □

Remark It follows from the definition of the map ψ in Lemma 5.3.5 that when $k \geq 2$ the lemma remains true if $W_{r,k}$ is replaced by the subset

$$\{w \in W_{r,k} : \text{ the first symbol in the string } w \text{ is } [\}.$$

Proof of Proposition 5.3.3 Let $W_{k,r}$ be as defined in Lemma 5.3.5, noting that the number of commutators of weight at most s in the letters $x_1, \ldots, x_r$ is at most

$$\sum_{k=1}^{s} |W_{k,r}|$$

by that lemma. We can specify a string belonging to $W_{k,r}$ by first saying which k symbols in the string are of the form x_i, and then by specifying which x_i each of these symbols is. This implies that

$$\begin{aligned} |W_{k,r}| &\leq r^k \binom{2k-1}{k} \\ &\leq r^s \binom{2s}{k}, \end{aligned}$$

and hence that

$$\begin{aligned} \sum_{k=1}^{s} |W_{k,r}| &\leq r^s \sum_{k=1}^{s} \binom{2s}{k} \\ &\leq r^s 2^{2s} \\ &= (4r)^s, \end{aligned}$$

which proves the proposition. □

Lemma 5.3.6 *Let α be a commutator in elements $x_1, \ldots, x_r$. Then when α is expressed as a word in the x_i and their inverses, each x_i and its inverse appear at most $2^{|\chi(\alpha)|-1}\chi_i(\alpha)$ times between them.*

Proof If $|\chi(\alpha)| = 1$ then the lemma is trivial, so we may assume that there exist commutators β_1, β_2 in the x_i such that $\alpha = [\beta_1, \beta_2]$ and, by induction, that when a given β_j is written as a word in the x_i and their inverses, each x_i and its inverse appear at most $2^{|\chi(\beta_j)|-1}\chi_i(\beta_j)$ times between them. It follows that when α is written as a word in the x_i and their inverses, each x_i and its inverse appear at most

$$2^{|\chi(\beta_1)|}\chi_i(\beta_1) + 2^{|\chi(\beta_2)|}\chi_i(\beta_2) \tag{5.3.3}$$

times between them. However, (5.3.3) is at most $2^{\max\{|\chi(\beta_1)|,|\chi(\beta_2)|\}}\chi_i(\alpha)$, so since $|\chi(\beta_j)| < |\chi(\alpha)|$ this implies the desired bound. □

5.4 The Collecting Process and Basic Commutators

In this section we explain how to generalise the expression (5.3.1) to an arbitrary nilpotent group. Expressing an element of the Heisenberg group in the form (5.3.1) was possible thanks to repeated use of the identity (5.1.4). In an arbitrary nilpotent group we use the following expanded set of commutator identities in an analogous way.

Lemma 5.4.1 *Let u and v be elements of an s-step nilpotent group. Then, denoting $v_0 = v$ and $v_{i+1} = [v_i, u]$ for $i > 1$, we have*

$$vu = uv[v,u], \tag{5.4.1}$$
$$v^{-1}u = u[v,u]^{-1}v^{-1}, \tag{5.4.2}$$
$$vu^{-1} = u^{-1}vv_2v_4\cdots v_5^{-1}v_3^{-1}v_1^{-1}, \tag{5.4.3}$$
$$v^{-1}u^{-1} = u^{-1}v_1v_3v_5\cdots v_4^{-1}v_2^{-1}v^{-1}, \tag{5.4.4}$$

the expressions on the right-hand side of identities (5.4.3) *and* (5.4.4) *being finite by Lemma 5.3.2.*

Proof This proof is partially based on Hall [38, §11.1]. Identities (5.4.1) and (5.4.2) are immediate. To prove (5.4.3) and (5.4.4) we use the commutator identity

$$[x, yz] = [x,z][x,y][x,y,z], \tag{5.4.5}$$

which is easily verified by direct computation. We first claim that

$$[v_k, u^{-1}] = v_{k+2}[v_{k+2}, u^{-1}]v_{k+1}^{-1} \tag{5.4.6}$$

for every $k \geq 0$. Indeed, (5.4.5) gives the commutator identity

$$1 = [x, yy^{-1}] = [x, y^{-1}][x,y][x,y,y^{-1}],$$

and hence the commutator identity

$$[x, y^{-1}] = [x,y,y^{-1}]^{-1}[x,y]^{-1}. \tag{5.4.7}$$

We then have

$$\begin{aligned}
[v_k, u^{-1}] &= [v_{k+1}, u^{-1}]^{-1}v_{k+1}^{-1} && \text{(by (5.4.7))}\\
&= \left([v_{k+2}, u^{-1}]^{-1}v_{k+2}^{-1}\right)^{-1} v_{k+1}^{-1} && \text{(by (5.4.7))}\\
&= v_{k+2}[v_{k+2}, u^{-1}]v_{k+1}^{-1},
\end{aligned}$$

giving (5.4.6) as claimed. Iterating (5.4.6) and noting that $v_k = 1$ for large enough k by Lemma 5.3.2, we deduce that

$$[v, u^{-1}] = v_2v_4\cdots v_5^{-1}v_3^{-1}v_1^{-1}. \tag{5.4.8}$$

Identities (5.4.3) and (5.4.4) then follow from substituting u^{-1} for u in (5.4.1) and (5.4.2) and applying (5.4.8). □

We apply the identities (5.4.1)–(5.4.4) to write a group element in a form analogous to (5.3.1) via an algorithm called the *collecting process.* Our description of the collecting process is based on Hall [38, §11.1].

Fix an s-step nilpotent group G and a generating set $x_1, \ldots, x_r$ for G. Write $c_1, \ldots, c_d$ for the set of commutators in the x_i of total weight at most s, ordered so that

(i) $c_i = x_i$ for $i = 1, \ldots, r$;
(ii) $|\chi(c_i)| \leq |\chi(c_j)|$ whenever $i \leq j$; and
(iii) commutators of the same weight vector are listed consecutively;

and arbitrarily otherwise. We fix this order from now on.

A string of commutators and their inverses $c_{i_1}^{\epsilon_1} \cdots c_{i_n}^{\epsilon_n}$ with each $\epsilon_j \in \{\pm 1\}$ is said to be in *collected form* if $i_1 \leq \cdots \leq i_n$. The collecting process is a process for converting a string that is not in collected form into a string that is in collected form, without changing the group element that the string represents.

Note, in particular, that the empty string, which we write as e, is in collected form and evaluates to the identity element $1 \in G$.

Given an arbitrary string $c = c_{i_1}^{\epsilon_1} \cdots c_{i_n}^{\epsilon_n}$ with each $\epsilon_j \in \{\pm 1\}$ such that c is not in collected form, set $m \in [n]_0$ maximal such that $c_{i_1}^{\epsilon_1} \cdots c_{i_m}^{\epsilon_m}$ is in collected form and such that $i_m \leq i_k$ for every $k = m+1, \ldots, n$, define $c_{i_1}^{\epsilon_1} \cdots c_{i_m}^{\epsilon_m}$ to be the *collected part* of the string c, and define $c_{i_{m+1}}^{\epsilon_{m+1}} \cdots c_{i_n}^{\epsilon_n}$ to be the *uncollected part.*

We define the collecting process by means of an operator $\mathcal{C}$ on the set of finite strings of commutators and their inverses as follows. If a string c is in collected form then we simply set $\mathcal{C}(c) = c$. If c is not in collected form then let u be the smallest subscript of a commutator appearing in the uncollected part of c, and let c_{i_j} be the left-most copy of c_u in the uncollected part. Next, define commutators $b_1, b_2, \ldots$ recursively by setting

$$b_1 = \begin{cases} [c_{i_{j-1}}, c_{i_j}] & \text{if } |\chi([c_{i_{j-1}}, c_{i_j}])| \leq s \\ e & \text{otherwise} \end{cases} \tag{5.4.9}$$

and

$$b_{k+1} = \begin{cases} [b_k, c_{i_j}] & \text{if } |\chi([b_k, c_{i_j}])| \leq s \\ e & \text{otherwise.} \end{cases} \tag{5.4.10}$$

Then, if $\epsilon_j = \epsilon_{j-1} = 1$, set

$$\mathcal{C}(c) = c_{i_1}^{\epsilon_1} \cdots c_{i_m}^{\epsilon_m} \cdots c_{i_j} c_{i_{j-1}} b_1 \cdots c_{i_n}^{\epsilon_n}$$

and say that $\mathcal{C}$ performs a *collecting transformation of type 1*, or simply a *transformation of type 1*, on c. If $\epsilon_j = 1$ and $\epsilon_{j-1} = -1$ then set

$$\mathcal{C}(c) = c_{i_1}^{\epsilon_1} \cdots c_{i_m}^{\epsilon_m} \cdots c_{i_j} b_1^{-1} c_{i_{j-1}}^{-1} \cdots c_{i_n}^{\epsilon_n}$$

and say that $\mathcal{C}$ performs a *(collecting) transformation of type 2* on c. If $\epsilon_j = -1$ and $\epsilon_{j-1} = 1$ then set

$$\mathcal{C}(c) = c_{i_1}^{\epsilon_1} \cdots c_{i_m}^{\epsilon_m} \cdots c_{i_j}^{-1} c_{i_{j-1}} b_2 b_4 \cdots b_5^{-1} b_3^{-1} b_1^{-1} \cdots c_{i_n}^{\epsilon_n}$$

and say that $\mathcal{C}$ performs a *(collecting) transformation of type 3* on c. Finally, if $\epsilon_j = \epsilon_{j-1} = -1$ then set

$$\mathcal{C}(c) = c_{i_1}^{\epsilon_1} \cdots c_{i_m}^{\epsilon_m} \cdots c_{i_j}^{-1} b_1 b_3 b_5 \cdots b_4^{-1} b_2^{-1} c_{i_{j-1}}^{-1} \cdots c_{i_n}^{\epsilon_n}$$

and say that $\mathcal{C}$ performs a *(collecting) transformation of type 4* on c.

Proposition 5.4.2 *Let G be an s-step nilpotent group generated by elements $x_1, \ldots, x_r$. Write $c_1, \ldots, c_d$ for the ordered list of commutators in the x_i of total weight at most s. Let $c = c_{i_1}^{\epsilon_1} \cdots c_{i_n}^{\epsilon_n}$ be a finite string of commutators and their inverses. Then there exists $k \in \mathbb{N}$ such that $\mathcal{C}^k(c)$ is in collected form, and $\mathcal{C}^k(c)$ evaluates to the same group element as c.*

Proof Let $w \geq 1$ be such that $i_\ell \geq w$ for every $\ell > m$. We proceed by downward induction on w, noting that the case $w = d$ is vacuously true since the uncollected part of c cannot possibly consist entirely of copies of c_d. We then induct on the number of copies of c_w appearing in the uncollected part of c. The proposition is trivial by the downward induction on w in the case where the number of copies of c_w appearing in the uncollected part of c is 0, so we may assume that there is at least one copy of c_w in the uncollected part of c. Note that this implies that $w = u$.

The operator $\mathcal{C}$ moves the commutator c_{i_j} to the $j-1$ position in the string c. It inserts some new commutators to the string, but the subscripts of all of these are greater than w. Thus, after $j - m + 1$ applications of $\mathcal{C}$, the commutator c_{i_j} will lie in the collected part of the string, and no new copies of any c_v with $v \leq w$ will have been inserted. This means that the smallest subscript of a commutator appearing in the uncollected part of $\mathcal{C}^{j-m+1}(c)$ is at least w, and that $\mathcal{C}^{j-m+1}(c)$ has fewer copies of c_w in its uncollected part than c does. It therefore follows by induction that there exists $k \in \mathbb{N}$ such that $\mathcal{C}^k(c)$ is in collected form, as required.

To see that $\mathcal{C}^k(c)$ evaluates to the same group element as c, note that the identities (5.4.1)–(5.4.4) and Lemma 5.3.2 imply that $\mathcal{C}(c)$ always evaluates to the same group element as c. □

It turns out that if we start with a finite string of the elements x_i and their inverses, with no higher-weight commutators, then only certain commutators will appear during the collecting process. For example, one will never need to interchange the positions of a copy of x_1 with a copy of $[x_3, x_2]$, since every copy of x_1 will have already been moved to the left-hand end of the string before any copies of $[x_3, x_2]$ arise, and so the commutator $[[x_3, x_2], x_1]$ will never arise. Those commutators that can arise are called *basic commutators.*

We can give an explicit recursive definition of the basic commutators, as follows.

Definition 5.4.3 (basic commutators) Let $x_1, \ldots, x_r$ be a finite set of letters. For $i = 1, \ldots, r$, set $u_i = x_i$ and declare the u_i to be the basic commutators of total weight 1. Then, having defined the basic commutators $u_1, \ldots, u_m$ of total weight less than k, define a commutator α of total weight k to be basic if

- $\alpha = [u_i, u_j]$ for some u_i, u_j with $i > j$; and
- if $u_i = [u_s, u_t]$ then $j \geq t$.

Then label the basic commutators of total weight k as $u_{m+1}, \ldots, u_{m'}$, in the same order that they appear in the ordered list of all commutators.

Note that the arbitrariness of the order implies that the list of basic commutators is not uniquely defined. Note, however, that if $r \geq 2$ the commutators $[[\cdots[[x_2, x_1], x_1]\cdots], x_1]$ are always basic, so there are always basic commutators of every total weight, whereas if $r = 1$ then x_1 is the unique basic commutator.

Lemma 5.4.4 *Let G be an s-step nilpotent group generated by elements $x_1, \ldots, x_r$, and let c be a finite string in the x_i and their inverses. Let $\ell \in \mathbb{N}$. Then $\mathcal{C}^\ell(c)$ consists entirely of basic commutators and their inverses.*

Proof Write $c_1, \ldots, c_d$ for the ordered list of commutators in the x_i of total weight at most s. The lemma holds by definition in the case $\ell = 1$. We may assume that $\ell \geq 2$ and write

$$\mathcal{C}^{\ell-1}(c) = c_{i_1}^{\epsilon_1} \cdots c_{i_n}^{\epsilon_n}.$$

We may also assume by induction that each c_{i_k} is a basic commutator.

If $\mathcal{C}^{\ell-1}(c)$ is in collected form then $\mathcal{C}^{\ell}(c) = \mathcal{C}^{\ell-1}(c)$ and the lemma holds by induction. If not then let u be the smallest subscript of a commutator appearing in the uncollected part of c, and let c_{i_j} be the leftmost copy of c_u in the uncollected part. Define commutators $b_1, b_2, \ldots$ as in the definition of $\mathcal{C}$. It is sufficient to show that the commutators b_k are all basic.

Since c_{i_j} and $c_{i_{j-1}}$ are basic, if b_1 is basic then so are $b_2, b_3, \ldots$. Since $i_{j-1} > i_j$, if $c_{i_{j-1}}$ is of weight 1 then b_1 is basic. We may therefore assume that $c_{i_{j-1}}$ is not of weight 1, say $c_{i_{j-1}} = [c_v, c_w]$. The commutator $c_{i_{j-1}}$ must have arisen from an earlier application of $\mathcal{C}$, in which a copy of the commutator c_w was moved to the left. At that point, w must have been the smallest subscript of a commutator appearing in the uncollected part of the string, which implies in particular that $w \leq u$. It follows that $b_1 = [[c_v, c_w], c_{i_j}] = [[c_v, c_w], c_u]$ is a basic commutator, as required. □

Theorem 5.4.5 ([38, Theorem 11.2.4]) *Let $s \in \mathbb{N}$. Let G be an s-step nilpotent group with generators $x_1, \ldots, x_r$, and let $u_1, \ldots, u_t$ be a complete ordered list of basic commutators of weight at most s in the x_i. Then every element of G can be expressed in the form*

$$u_1^{\ell_1} \cdots u_t^{\ell_t} \tag{5.4.11}$$

with $\ell_i \in \mathbb{Z}$.

Proof An arbitrary element $g \in G$ can by definition be written as a string in the elements x_i and their inverses, and hence as a string c in the commutators $c_1, \ldots, c_r$ and their inverses. Proposition 5.4.2 implies that there exists $k \in \mathbb{N}$ such that $\mathcal{C}^k(c)$ is in collected form, and that $\mathcal{C}^k(c) = g$ in G. Since each of $c_1, \ldots, c_r$ is a basic commutator by definition, the string $\mathcal{C}^k(c)$ consists entirely of basic commutators by Lemma 5.4.4. A string in collected form consisting entirely of basic commutators is automatically of the form (5.4.11), and so the theorem is proved. □

Lemma 5.4.6 ([69, Lemma 3.6]) *Let $r \in \mathbb{N}$ and let $j \in [r]$. Let $x_1, \ldots, x_r$ be letters, and let $u_1, u_2, \ldots$ be a complete list of basic commutators in the x_i. Then there exists a complete list of basic commutators in the letters $x_1, \ldots, x_{j-1}, x_{j+1}, \ldots, x_r$ that is precisely the subsequence of those u_i with zero x_j-weight.*

Proof The ordered list of basic commutators of weight 1 is $x_1, \ldots, x_{j-1}, x_{j+1}, \ldots, x_r$. When defining the basic commutators of weight n we may therefore assume by induction that the sequence of basic commutators of weight less than n is precisely the subsequence of those u_i of total weight less than n and zero x_j-weight. It is then trivial that a commutator of weight exactly n with zero x_j-weight satisfies the conditions given in Definition 5.4.3 for being included as a basic commutator in one list, if and only if it satisfies the conditions for inclusion in the other list. If we then choose the order of the basic commutators of total weight n in $x_1, \ldots, x_{j-1}, x_{j+1}, \ldots, x_r$ to be the restriction of the order on the u_i, it follows that the sequence of basic commutators of weight at most n in $x_1, \ldots, x_{j-1}, x_{j+1}, \ldots, x_r$ is precisely the subsequence of those u_i of total weight at most n and zero x_j-weight, as required. □

Corollary 5.4.7 *Let $s \in \mathbb{N}$. Let G be an s-step nilpotent group with generators $x_1, \ldots, x_r$, and let $y \in \langle x_{i_1}, \ldots, x_{i_m} \rangle$. Then when y is expressed in the form $u_1^{\ell_1} \cdots u_t^{\ell_t}$ given by Theorem 5.4.5, we may assume that $\ell_k = 0$ for every k for which there exists $j \notin \{i_1, \ldots, i_m\}$ with $\chi_j(u_k) > 0$.*

Before we move on, it is worth saying a word about the motivation for the term *basic commutators*. This entails a definition: if F is the free group on generators $u_1, \ldots, u_r$, we define the *free s-step nilpotent group of rank r*, denoted $N_{r,s}$, to be the quotient F/F_{s+1}. More precisely, writing $x_i = u_i F_{s+1} \in N_{r,s}$ for each i, we say that $N_{r,s}$ is the *free s-step nilpotent group on generators $x_1, \ldots, x_r$*. The importance of free nilpotent groups stems at least in part from the fact that if G is a nilpotent group of step at most s generated by elements $y_1, \ldots, y_r$ then there exists a homomorphism $\varphi : N_{r,s} \to G$ such that $\varphi(x_i) = y_i$ for $i = 1, \ldots, r$.

The term *basic commutators* refers to the fact that the expression (5.4.5) for an element of $G = N_{r,s}$ given by Theorem 5.4.5 is unique. Indeed, for every $k = 1, \ldots, s$ the basic commutators of weight k form a basis of the free abelian group G_k/G_{k+1}. A proof of this is beyond the scope of this book, but can be found in Hall [38, Theorem 11.2.4].

5.5 Commutator Forms

Our main aim in the rest of this chapter is to prove results like Proposition 5.1.3 and Lemma 5.1.4 for arbitrary nilpotent groups. A property of commutators in the Heisenberg group H that was convenient in the proof of those results was (5.1.7), which says that the 'commutator map'

$$\begin{array}{cccc} [\ ,\ ] & : & \langle u_1 \rangle \times \langle u_2 \rangle & \to & [H,H] \\ & & (x,y) & \mapsto & [x,y] \end{array}$$

is a homomorphism in each variable. In this section we generalise this property to arbitrary nilpotent groups.

The first step is to give a formal means of describing more general 'commutator maps'. The definition is perhaps easier to understand by means of an example: given a group G and a commutator $\alpha = [[x_1, x_2], x_1]$, we define the *commutator map on G associated to α*, denoted by $\alpha_G : G \times G \to G$, via $\alpha_G(g_1, g_2) = [[g_1, g_2], g_1]$. Formally, we use the following definition, which is based on [72, Definition 3.2].

Definition 5.5.1 (commutator forms) Let $x_1, \ldots, x_r$ be letters. Then given a group G and a commutator α in the letters x_i we denote by $\alpha_G : G^r \to G$ the *commutator map on G associated to α*, which we define by setting $\alpha_G(g_1, \ldots, g_r)$ to be the element of G obtained by first replacing each x_i in the commutator α by g_i to obtain a commutator in the g_i, and then evaluating that commutator in G using the rule (5.3.2). We call the class of commutator maps associated to a given commutator a *commutator form*. By abuse of notation we often denote both the commutator form and each commutator map in it simply by α.

We define the *weight vector*, *x_i-weight*, *total weight* or *degree* of a commutator form or commutator map to be the same as that of the corresponding commutator. We also define a commutator form or commutator map to be *basic*, *linear* or *trivial* if the corresponding commutator is.

Throughout this book we make repeated use of the following fundamental fact. The proof we give essentially appears in [3, §6], for example.

Lemma 5.5.2 *Let G be a group, and let $\alpha : G \times \cdots \times G \to G_s$ be a linear commutator map of weight s. Then α is a homomorphism modulo G_{s+1} in each variable. Moreover, $[G,G]$ is in the kernel of each of these homomorphisms.*

The $s = 2$ case of Lemma 5.5.2 is in fact a special case of the following more general fact.

Lemma 5.5.3 *Let G be a group and let A, B be subgroups of G. If $[A, B]$ commutes with A then the commutator map*

$$\begin{aligned}[\ ,\ ] : A \times B &\to [A, B] \\ (a, b) &\mapsto [a, b]\end{aligned}$$

is a homomorphism in the first variable; if $[A, B]$ commutes with B then it is a homomorphism in the second variable. In particular, by Proposition 5.2.7, for every $i, j \in \mathbb{N}$ we have

$$[\ ,\ ] : G_i \times G_j \to G_{i+j},$$

and this map is a homomorphism modulo G_{i+j+1} in each variable.

Proof We prove that if $[A, B]$ commutes with A then $[\ ,\ ] : A \times B \to [A, B]$ is a homomorphism in the first variable; the proof for the second variable is similar. Let $a_1, a_2 \in A$ and $b \in B$. The identity (5.1.4) gives

$$a_1a_2b = ba_1a_2[a_1a_2, b] \tag{5.5.1}$$

and

$$a_1a_2b = ba_1[a_1, b]a_2[a_2, b]. \tag{5.5.2}$$

Since $[a_1, b]$ commutes with a_2, (5.5.2) implies that

$$a_1a_2b = ba_1a_2[a_1, b][a_2, b], \tag{5.5.3}$$

and then comparing (5.5.1) and (5.5.3) gives $[a_1a_2, b] = [a_1, b][a_2, b]$, as required. □

Proof of Lemma 5.5.2 Let α be a commutator form of weight s. The lemma is trivial if $s = 1$, so we may assume that $s > 1$, and hence that there exist commutator forms β_1 and β_2 such that $\alpha = [\beta_1, \beta_2]$. Lemma 5.5.3 implies that the map $[\ ,\ ] : G_{|\chi(\beta_1)|} \times G_{|\chi(\beta_2)|} \to G_s$ is a homomorphism modulo G_{s+1} in each variable. Proposition 5.2.7 also implies that the kernel of each of these homomorphisms in the first variable contains $G_{|\chi(\beta_1)|+1}$, and the kernel of each of these maps in the second variable contains $G_{|\chi(\beta_2)|+1}$. By induction on s, for each i the commutator form $\beta_i : G^{|\chi(\beta_i)|} \to G_{|\chi(\beta_i)|}$ is a homomorphism modulo $G_{|\chi(\beta_i)|+1}$ in each variable, and so we may conclude that α is a homomorphism modulo G_{s+1} in each variable, as required. The fact that $[G, G]$ is in the kernel of each of these homomorphisms follows from Proposition 5.2.7. □

5.6 Nilprogressions

The 'progression' $P(x; L)$ of Proposition 5.1.3 is in fact an instance of the following more general definition.

Definition 5.6.1 (nilprogression) Let $x_1, \ldots, x_r$ be elements in a group G, and let $L_1, \ldots, L_r \geq 0$. Then the *non-abelian progression* $P(x; L) = P(x_1, \ldots, x_r; L_1, \ldots, L_r)$ is defined to consist of all those elements of G that can be expressed as words in the x_i and their inverses in which each x_i and its inverse appear at most L_i times between them. We define r to be the *rank* of $P(x; L)$. If the x_i generate an s-step nilpotent group then $P(x; L)$ is said to be a *nilprogression* of step s, and in this instance we often write $P_{\text{nil}}(x; L)$ instead of $P(x; L)$.

We can also generalise the 'progression' $\overline{P}(x; L)$ defined in Lemma 5.1.4, as follows. From now on, given $L_1, \ldots, L_r \geq 0$ and a vector $\chi \in \mathbb{N}_0^r$ we abbreviate by L^χ the quantity $L_1^{\chi_1} \cdots L_r^{\chi_r}$.

Definition 5.6.2 (nilpotent progression; [12, Definition 1.4]) Let G be a nilpotent group. Let $x_1, \ldots, x_r$ be elements of G, and let $u_1, \ldots, u_t$ be the list of basic commutators in the x_i. Let $L_1, \ldots, L_r \geq 0$. Then the *nilpotent progression* $\overline{P}(x; L) = \overline{P}(x_1, \ldots, x_r; L_1, \ldots, L_r)$ is defined to be the set

$$\overline{P}(x; L) = \{u_1^{\ell_1} \cdots u_t^{\ell_t} : |\ell_i| \leq L^{\chi(u_i)}\}.$$

We define r to be the *rank* of $\overline{P}(x; L)$.

The first main aim of this section is to show that nilprogressions and nilpotent progressions have small tripling, as follows.

Proposition 5.6.3 *Fix $r, s \geq 1$, and let $x_1, \ldots, x_r$ be elements in an s-step nilpotent group. Then for every $L_1, \ldots, L_r \in \mathbb{N}$ the nilpotent progression $\overline{P}(x; L)$ satisfies $|\overline{P}(x; L)^3| \ll_{r,s} |\overline{P}(x; L)|$. Moreover, there is a constant $\lambda_{r,s}$ depending only on r and s such that if $L_1, \ldots, L_r \geq \lambda_{r,s}$ then the nilprogression $P_{\text{nil}}(x; L)$ also satisfies $|P_{\text{nil}}(x; L)^3| \ll_{r,s} |P_{\text{nil}}(u; L)|$.*

We invite the reader to show in Exercise 5.5 that the assumption that $L_1, \ldots, L_r \geq \lambda_{r,s}$ is necessary in the last part of Proposition 5.6.3.

The second main aim of this section is to show that nilprogressions, nilpotent progressions and ordered progressions in nilpotent groups are roughly equivalent, in the following sense.

Proposition 5.6.4 *Let G be an s-step nilpotent group, let $x_1, \ldots, x_r \in G$, and let $L_1, \ldots, L_r \in \mathbb{N}$. Then*

$$P_{\mathrm{ord}}(x; L) \subset P_{\mathrm{nil}}(x; L) \subset \overline{P}(x; L) \subset P_{\mathrm{ord}}(x; L)^{(96s)^{s^2} r^s}.$$

Proposition 5.6.3 actually follows fairly straightforwardly from Proposition 5.6.4, as follows.

Proof of Proposition 5.6.3 from Proposition 5.6.4 It is trivial from the definitions that

$$|\overline{P}(x; CL)| \leq C^{O_{r,s}(1)} |\overline{P}(x; L)| \tag{5.6.1}$$

for every $C \geq 1$ and

$$P_{\mathrm{nil}}(x; L)^m \subset P_{\mathrm{nil}}(x; mL) \tag{5.6.2}$$

for every $m \in \mathbb{N}$. Moreover, (5.6.2) combines with Proposition 5.6.4 to imply that

$$\overline{P}(x; L) \subset P_{\mathrm{nil}}(x; O_{r,s}(L)). \tag{5.6.3}$$

To prove the first assertion of the proposition, we write

$$\begin{aligned} \overline{P}(x; L)^3 &\subset P_{\mathrm{nil}}(x; O_{r,s}(L)) && \text{(by (5.6.3))} \\ &\subset \overline{P}(x; O_{r,s}(L)) && \text{(by Proposition 5.6.4)} \end{aligned}$$

and then apply (5.6.1). To prove the second assertion, we first note that (5.6.3) implies that there exists c depending only on r and s such that $\overline{P}(x; \lfloor cL \rfloor) \subset P_{\mathrm{nil}}(x; L)$; if each L_i satisfies $L_i \geq 1/c$, this implies in particular that

$$\overline{P}(x; \tfrac{1}{2}cL) \subset P_{\mathrm{nil}}(x; L).$$

On the other hand, we have

$$P_{\mathrm{nil}}(x; L)^3 \subset \overline{P}(x; 3L)$$

by (5.6.2) and Proposition 5.6.4. These last two containments combine with (5.6.1) to imply the second assertion of the proposition. □

The first inclusion of Proposition 5.6.4 is trivial. We prove the second in the following lemma, essentially just by keeping track of the commutators arising during the collecting process.

Lemma 5.6.5 ([72, Corollary A.8]) *Let G be an s-step nilpotent group, let $x_1, \ldots, x_r \in G$, and let $L_1, \ldots, L_r \in \mathbb{N}$. Then $P_{\mathrm{nil}}(x; L) \subset \overline{P}(x; L)$.*

The final inclusion of Proposition 5.6.4 follows from Proposition 5.3.3 and the following result.

Proposition 5.6.6 *Let $x_1, \dots, x_r$ be elements in an s-step nilpotent group G. Let $L_1, \dots, L_r \in \mathbb{N}$. Let α be a basic commutator form of weight at most s on r variables. Then for every integer ℓ satisfying $|\ell| \le L^{\chi(\alpha)}$ we have*

$$\alpha(x_1, \dots, x_r)^\ell \in P_{\mathrm{ord}}(x; L)^{(24s)^{s(s+1-|\chi(\alpha)|)}}.$$

The constants 24 in Proposition 5.6.6 and 96 in Proposition 5.6.4 could be improved with a little more care, but we are more concerned with the shape of the bounds than with the precise constants involved.

We start by proving Lemma 5.6.5, which as it happens is also an ingredient in Proposition 5.6.6.

Proof of Lemma 5.6.5 We follow the proof of [72, Lemma A.6]. Write $u_1, \dots, u_t$ for a complete ordered list of basic commutators of weight at most s in the x_i. We claim that if w is a word in the x_i and their inverses featuring p_j copies of x_j and n_j copies of x_j^{-1}, with $p_j + n_j \le L_j$, then applying the collecting process to w results in at most $L^{\chi(u_i)}$ copies of u_i and u_i^{-1} between them. This proves the lemma.

The collecting process never creates any new copies of x_i or x_i^{-1}, so the claim certainly holds whenever u_i has total weight 1. We may therefore, by induction, prove the claim for basic commutators of a given total weight $\omega > 1$, under the assumption that it holds for all commutators of total weight less than ω.

Since $\omega > 1$, any commutator of weight ω arising from the collecting process is of the form $[u_i, u_j]^{\pm 1}$ for some basic commutators u_i, u_j of total weight less than ω. We prove the claim by defining an injection f from the set of copies of $[u_i, u_j]^{\pm 1}$ to the set of pairs of copies of $u_i^{\pm 1}$ and $u_j^{\pm 1}$. This is sufficient as by induction the number of such pairs is at most $L^{\chi(u_i)} L^{\chi(u_j)} = L^{\chi([u_i, u_j])}$.

If a copy z of $[u_i, u_j]^{\pm 1}$ arose from a collecting transformation of type 1 or 2 then this can only have been as a result of interchanging a copy a of $u_i^{\pm 1}$ and a copy b of u_j. In this case simply define $f(z) = (a, b)$.

If z arose as from a collecting transformation of type 3 or 4 then it must have arisen from interchanging a copy b' of u_j^{-1} and a copy of some commutator u_k. Defining $v_0, v_1, v_2, \dots$ by setting $v_0 = u_k$ and $v_{q+1} = [v_q, u_j]$, it must therefore be the case that u_i is equal to some v_q. If $q = 0$ then z arose from interchanging a copy a' of u_i with b', and we define $f(z) = (a', b')$. If $q > 0$ then exactly one copy a'' of v_q will also have arisen as a result of the same transformation as that producing z. In this case we define $f(z) = (a'', b')$.

Now that we have defined f it remains to show that it is an injection. A given pair of copies of $u_i^{\pm 1}$ and $u_j^{\pm 1}$ will be interchanged at most once during the collection process, and so if two distinct copies z, z' of $[u_i, u_j]$ both arose from interchanging copies of u_i and u_j (which is to say as a result of a transformation of type 1 or 2, or of type 3 or 4 in the case $q = 0$) then we certainly have $f(z) \neq f(z')$. Furthermore, a given transformation of type 3 or 4 produces at most one copy of any commutator u_l, and so if z and z' both arose from transformations of type 3 or 4 in the case $q > 0$ then we also have $f(z) \neq f(z')$.

Finally, suppose a copy z of $[u_i, u_j]^{\pm 1}$ arose from a transformation of type 3 or 4 in the case $q > 0$. By definition of these transformations, the copy a'' of $u_i = v_q$ produced as a result of the transformation producing z will already be to the right of b', and so a'' and b' will never have to be interchanged and so no z' arising from a transformation of type 1 or 2, or of type 3 or 4 in the case $q = 0$, will have $f(z') = (a'', b')$. □

We now move on to the proof of Proposition 5.6.6. The following lemma was applied in a similar context by Breuillard and Green [12].

Lemma 5.6.7 *Let $L_1, \ldots, L_k \in \mathbb{N}$. Then every natural number $\ell \leq L_1 \cdots L_k$ can be written as the sum of at most k numbers of the form $\ell_1 \cdots \ell_k$ with $\ell_i \in \mathbb{N}$ and $\ell_i \leq L_i$.*

Proof The case $k = 1$ is trivial. When $k > 1$ we may write $\ell = mL_1 \cdots L_{k-1} + r$ for some $m, r \in \mathbb{N}$ with $m \leq L_k$ and $r \leq L_1 \cdots L_{k-1}$. By induction the number r can be written as the sum of at most $k - 1$ numbers of the form $\ell_1 \cdots \ell_{k-1}$, and so the lemma is proved. □

Lemma 5.6.7 allows us to reduce Proposition 5.6.6 to the case in which $\ell = \ell_1 \cdots \ell_r$ with $|\ell_i| \leq L_i$ for each i. If α is linear, Lemma 5.5.2 then allows us to replace $\alpha(x_1, \ldots, x_r)^\ell$ by $\alpha(x_1^{\ell_1}, \ldots, x_r^{\ell_r})$, at least modulo $G_{|\chi(\alpha)|}$. This is useful because studying $\alpha(x_1^{\ell_1}, \ldots, x_r^{\ell_r})$ is often more convenient than studying $\alpha(x_1, \ldots, x_r)^\ell$, not least thanks to the following observation.

Lemma 5.6.8 *Let α be a linear commutator form on r variables. Let $x_1, \ldots, x_r$ be elements of a nilpotent group G. Let $L_1, \ldots, L_r \in \mathbb{N}$. Then for all integers $\ell_1, \ldots, \ell_r$ satisfying $|\ell_i| \leq L_i$ we have $\alpha(x_1^{\ell_1}, \ldots, x_r^{\ell_r}) \in \overline{P}(x; 2^{|\chi(\alpha)|-1}L)$.*

Proof Since α is linear we have $\chi_i(\alpha) \leq 1$ for every i, so Lemma 5.3.6 implies that $\alpha(x_1^{\ell_1}, \ldots, x_r^{\ell_r}) \in P_{\text{nil}}(x; 2^{|\chi(\alpha)|-1}L)$. The lemma then follows from Lemma 5.6.5. □

Of course, we need to prove Proposition 5.6.6 for commutators of arbitrary degree, not just linear ones. To get around this we introduce a definition. If a commutator α is not linear then we define its *linearisation* α' to be the commutator on letters $x_{1,1}, \ldots, x_{1,\chi_1(\alpha)}, \ldots, x_{r,1}, \ldots, x_{r,\chi_r(\alpha)}$ obtained by replacing each instance of x_i in α by a different $x_{i,j}$. Thus, for example, the linearisation of $[[x_2, x_1], x_1]$ is $[x_{2,1}, x_{1,1}], x_{1,2}]$. The linearisation of the commutator form associated to α is then the commutator form associated to α'. Given a linearised commutator form α', we abbreviate

$$\alpha'(x_{i,j}^{\ell_{i,j}}) = \alpha'(x_{1,1}^{\ell_{1,1}}, \ldots, x_{1,\chi_1(\alpha)}^{\ell_{1,\chi_1(\alpha)}}, \ldots, x_{r,1}^{\ell_{r,1}}, \ldots, x_{r,\chi_r(\alpha)}^{\ell_{r,\chi_r(\alpha)}})$$

for $\ell_{i,j} \in \mathbb{Z}$.

Proof of Proposition 5.6.6 For each $i = 1, \ldots, r$, set $x_{i,j} = x_i$ for $j = 1, \ldots, |\chi_i(\alpha)|$, so that if α' is the linearisation of α then

$$\alpha(x_1, \ldots, x_r) = \alpha'(x_{1,1}, \ldots, x_{1,\chi_1(\alpha)}, \ldots, x_{r,1}, \ldots, x_{r,\chi_r(\alpha)}).$$

We first consider powers of α of the form

$$\alpha(x_1, \ldots, x_r)^{\prod_{i=1}^r \prod_{j=1}^{\chi_j(\alpha)} \ell_{i,j}}$$

with $\ell_{i,j} \in \mathbb{Z}$ and $|\ell_{i,j}| \leq L_i$. For such $\ell_{i,j}$, it follows from Lemma 5.5.2 that

$$\alpha(x_1, \ldots, x_r)^{\prod_{i=1}^r \prod_{j=1}^{\chi_j(\alpha)} \ell_{i,j}} \in \alpha'(x_{i,j}^{\ell_{i,j}}) G_{|\chi(\alpha)|+1}, \tag{5.6.4}$$

and from Lemma 5.3.6 that

$$\alpha'(x_{i,j}^{\ell_{i,j}}) \in P_{\text{ord}}(x; L)^{2^{s-1}s}. \tag{5.6.5}$$

If $|\chi(\alpha)| = s$ then (5.6.4) becomes simply $\alpha(x_1, \ldots, x_r)^{\prod_{i=1}^r \prod_{j=1}^{\chi_j(\alpha)} \ell_{i,j}} = \alpha'(x_{i,j}^{\ell_{i,j}})$ by Corollary 5.2.9. Moreover, Lemma 5.6.7 implies that the element $\alpha(x_1, \ldots, x_r)^\ell$ can be written as the product of at most s elements of the form

$$\alpha(x_1, \ldots, x_r)^{\prod_{i=1}^r \prod_{j=1}^{\chi_j(\alpha)} \ell_{i,j}}$$

with $|\ell_{i,j}| \leq L_i$. When $|\chi(\alpha)| = s$ the proposition therefore follows from (5.6.4) and (5.6.5), so we may assume from now on that

$$|\chi(\alpha)| < s \tag{5.6.6}$$

and, by induction, that the proposition holds for every basic commutator form of total weight greater than $|\chi(\alpha)|$.

The containment (5.6.5) implies in particular that

$$\alpha'(x_{i,j}^{\ell_{i,j}}) \in P_{\text{nil}}(x; 2^{s-1}sL),$$

which combines with Lemma 5.6.5 to give

$$\alpha'(x_{i,j}^{\ell_{i,j}}) \in \overline{P}(x; 2^{s-1}sL). \tag{5.6.7}$$

Write $\beta_1, \ldots, \beta_k$ for the ordered list of those basic commutators in the x_i that have total weights greater than $|\chi(\alpha)|$ and at most s and satisfy $\chi_i(\beta_j) = 0$ for every i with $\chi_i(\alpha) = 0$. It follows from (5.6.4), (5.6.7), Theorem 5.4.5 and Corollary 5.4.7 that

$$\alpha'(x_{i,j}^{\ell_{i,j}}) = \alpha(x_1, \ldots, x_r)^{\prod_{i=1}^r \prod_{j=1}^{\chi_j(\alpha)} \ell_{i,j}} \beta_1^{m_1} \cdots \beta_k^{m_k}$$

for some $m_1, \ldots, m_k \in \mathbb{Z}$ with $|m_i| \leq (2^{s-1}sL)^{\chi(\beta_i)}$ for each i, and then from (5.6.5) that

$$\alpha(x_1, \ldots, x_r)^{\prod_{i=1}^r \prod_{j=1}^{\chi_j(\alpha)} \ell_{i,j}} \in P_{\text{ord}}(x; L)^{2^{s-1}s} \beta_k^{-m_k} \cdots \beta_1^{-m_1}. \tag{5.6.8}$$

Writing $x'_1, \ldots, x'_q$ for those x_i with $\chi_i(\alpha) = 0$, and $L'_1, \ldots, L'_q$ for the corresponding L_i, it follows from Lemma 5.4.6 and the definition of the β_j that each β_j is a basic commutator in the x'_i. The induction hypothesis therefore implies that $\beta_j^{m_j} \in P_{\text{ord}}(x'; 2^{s-1}sL')^{(24s)^{s(s-|\chi(\alpha)|)}}$ for each j, and so since $P_{\text{ord}}(x'; 2^{s-1}sL') \subset P_{\text{ord}}(x'; L')^{2^{s-1}sq}$ we have

$$\begin{aligned} \beta_j^{m_j} &\in P_{\text{ord}}(x'; L')^{2^{s-1}sq(24s)^{s(s-|\chi(\alpha)|)}} \\ &\subset P_{\text{ord}}(x; L)^{2^{s-1}sq(24s)^{s(s-|\chi(\alpha)|)}}. \end{aligned}$$

It therefore follows from (5.6.8) that

$$\begin{aligned} &\alpha(x_1, \ldots, x_r)^{\prod_{i=1}^r \prod_{j=1}^{\chi_j(\alpha)} \ell_{i,j}} \\ &\qquad \in P_{\text{ord}}(x; L)^{2^{s-1}s} P_{\text{ord}}(x; L)^{2^{s-1}skq(24s)^{s(s-|\chi(\alpha)|)}}. \end{aligned}$$

Since $q \leq |\chi(\alpha)|$, and since Proposition 5.3.3 and Lemma 5.4.6 imply that $k \leq (4|\chi(\alpha)|)^s$, applying (5.6.6) therefore gives

$$\begin{aligned} &\alpha(x_1, \ldots, x_r)^{\prod_{i=1}^r \prod_{j=1}^{\chi_j(\alpha)} \ell_{i,j}} \\ &\qquad \in P_{\text{ord}}(x; L)^{2^{s-1}s((4(s-1))^s(s-1)(24s)^{s(s-|\chi(\alpha)|)}+1)} \\ &\qquad \subset P_{\text{ord}}(x; L)^{2^{s-1}s(4s)^s s(24s)^{s(s-|\chi(\alpha)|)}}. \end{aligned}$$

Lemma 5.6.7 implies that $\alpha(x_1, \ldots, x_r)^\ell$ can be written as the product of at most s elements of the form $\alpha(x_1,\ldots,x_r)^{\prod_{i=1}^r \prod_{j=1}^{\chi_j(\alpha)} \ell_{i,j}}$ with $|\ell_{i,j}| \leq L_i$, so it follows that

$$\alpha(x_1,\ldots,x_r)^\ell \in P_{\mathrm{ord}}(x;L)^{2^{s-1}s^3(4s)^s(24s)^{s(s-|\chi(\alpha)|)}}.$$

Applying the crude bound $s^3 \leq 3^s$ for $s \in \mathbb{N}$, we deduce that

$$\begin{aligned}\alpha(x_1,\ldots,x_r)^\ell &\in P_{\mathrm{ord}}(x;L)^{(24s)^s(24s)^{s(s-|\chi(\alpha)|)}}\\ &= P_{\mathrm{ord}}(x;L)^{(24s)^{s(s+1-|\chi(\alpha)|)}},\end{aligned}$$

as required. □

Exercises

5.1 Let A be a subset of a group G, let H be another group, and suppose that $\varphi : A \to H$ is a Freiman 2-homomorphism. Show that if G is abelian then so is $\langle \varphi(A) \rangle$. Show, however, that if G is s-step nilpotent with $s \geq 2$ then $\langle \varphi(A) \rangle$ need not be. Formulate a stronger condition on φ (weaker than φ being a group homomorphism) that does force $\langle \varphi(A) \rangle$ to be s-step nilpotent if G is.

5.2 Given a group G and $m \in \mathbb{N}$, define $G^m = \langle g^m : g \in G \rangle$ to be the subgroup of G generated by the mth powers of elements of G.

(a) If G is nilpotent of step s, show that $(G^m)_s = (G_s)^{m^s}$.

(b) If G is nilpotent of step s and has rank at most r, show that G^m has index at most $m^{O_{r,s}(1)}$ in G.

5.3 Let G be an s-step nilpotent group, and let $A \subset G$ be a finite K-approximate group in which every element has order at most r. Show that there exists a subgroup H of G of size at most $r^{sK^s}|A|$ such that $A \subset H$.

5.4 Show directly that the nilprogression $P(x;L)$ defined in Proposition 5.1.3 is an $O(1)$-approximate group. That is, given arbitrary $L_1, L_2 \in \mathbb{N}$, give an explicit set X whose cardinality is bounded independently of L_1, L_2 such that $P(x;L)^2 \subset XP(x;L)$.

5.5 Give examples to show that a nilprogression

$$P_{\mathrm{nil}}(x_1,\ldots,x_r;L_1,\ldots,L_r)$$

of step s does not necessarily have small tripling if one of the L_i is bounded above in terms of r and s.

5.6 Let $x_1, \ldots, x_r$ be elements of a group G, and let $L_1, \ldots, L_r \geq 0$. Suppose that there exists some C such that whenever $1 \leq i < j \leq r$ we have

$$[x_i^{\pm 1}, x_j^{\pm 1}] \in P_{\mathrm{ord}}\left(x_{j+1}, \ldots, x_r; \frac{CL_{j+1}}{L_i L_j}, \ldots, \frac{CL_r}{L_i L_j}\right).$$

Show that $\langle x_1, \ldots, x_r \rangle$ is nilpotent of step at most $r-1$ and that $|P_{\mathrm{ord}}(x; L)^3| \ll_{C,r,s} |P_{\mathrm{ord}}(x; L)|$.

5.7 Let G be an s-step nilpotent group with generators $x_1, \ldots, x_r$, and let $u_1, \ldots u_t$ be a complete ordered list of basic commutators of weight at most s in the x_i. Define a partial order on $\mathbb{N}_0^r$, and hence on all weight vectors of commutators in the x_i, by writing $\chi \geq \chi'$ if $\chi_i \geq \chi'_i$ for every i. Let β be a commutator in the x_i. Show that there exist non-negative integers $\ell_1, \ldots, \ell_t$, such that $\chi(u_i) \geq \chi(\beta)$ whenever $\ell_i \neq 0$, and such that $\beta = u_1^{\ell_1} \cdots u_t^{\ell_t}$.

5.8 Let G be an s-step nilpotent group, let $x_1, \ldots, x_r \in G$, and let $u_1, \ldots u_t$ be a complete ordered list of basic commutators of weight at most s in the x_i. Given $L_1, \ldots, L_r > 0$, by definition we have

$$\overline{P}(x; L) = P_{\mathrm{ord}}(u_1, \ldots, u_t; L^{\chi(u_1)}, \ldots, L^{\chi(u_t)}).$$

Deduce from Exercise 5.7 that there exists $C \ll_{r,s} 1$ such that the elements $u_1, \ldots, u_t$ and positive real numbers $L^{\chi(u_1)}, \ldots, L^{\chi(u_t)}$ satisfy the hypothesis of Exercise 5.6. Use this to give a different proof of the conclusion of Proposition 5.6.3 that $|\overline{P}(x; L)^3| \ll_{r,s} |\overline{P}(x; L)|$.

6
Nilpotent Approximate Groups

6.1 Introduction and Overview of the Torsion-Free Case

In this chapter we prove the following nilpotent analogue of the Freiman–Green–Ruzsa theorem, due to the author.

Theorem 6.1.1 ([72, Theorem 1.5]) *Let G be a nilpotent group of step at most s, and suppose that $A \subset G$ is a finite K-approximate group. Then there exist a subgroup H of G normalised by A, a natural number $r \leq O_s(K^{e^{O(s)}})$, elements $x_1, \dots, x_r \in G$, and natural numbers $L_1, \dots, L_r$ such that*

$$A \subset HP_{\text{nil}}(x; L) \subset H\overline{P}(x; L) \subset A^{K^{e^{O(s)}}}.$$

In particular, $|H\overline{P}(x; L)| \leq \exp(K^{e^{O(s)}})|A|$ by Proposition 2.5.3.

Remarks Note that for fixed s the bounds $K^{e^{O(s)}}$ in the conclusion of Theorem 6.1.1 are polynomial in K. In Chapter 8 we prove a version of Theorem 6.1.1 in which the bounds are completely independent of s, though no longer polynomial in K.

The author has recently shown that the bounds in Theorem 6.1.1 can be improved if one assumes the bounds in Theorem 4.1.3 obtained by Sanders and stated in Exercise 4.4 [74]. For example, the bound on r in Theorem 6.1.1 can be improved to $e^{O(s^2)}K \log^{O(s)} K$. We guide the reader to these improved bounds in Exercise 6.1.

Remark 6.1.2 It is worth highlighting at this point that if $K < 2$ then every K-approximate group is an exact group, and so the conclusion of Theorem 6.1.1 holds trivially. In proving Theorem 6.1.1, we may therefore assume that $K \geq 2$, which confers the minor notational advantage of being able to absorb multiplicative constants of K into powers. For example, it is easy to check (and we encourage the reader to do so) that if $K \geq 2$ then $O(K^{O(1)}) \leq K^{O(1)}$. We absorb multiplicative constants in this way throughout this chapter, and indeed the rest of the book, often without further comment.

The proof of Theorem 6.1.1 in the case where G has no torsion is fairly quick, and yet still features many of the key ideas from the general proof. We therefore start by restricting to that case. Theorem 6.1.1 is essentially due to Breuillard and Green [12] in the torsion-free case, although the proof we give here is completely different to theirs.

The basic strategy of the proof of Theorem 6.1.1 is to split a set A of small tripling in a nilpotent group into sets of small tripling that generate subgroups of lower step, and then to induct on the step. In the torsion-free case we have the following particularly clean statement.

Proposition 6.1.3 *Let $K, s \geq 2$, let G be a torsion-free s-step nilpotent group, and suppose that $A \subset G$ is a finite K-approximate group. Then there exist $k \leq K^{O(1)}$ and finite $K^{O(1)}$-approximate groups*

$$A_1, \dots, A_k \subset A^{O(1)},$$

each of which generates a subgroup of step less than s, such that $A \subset A_1 \cdots A_k$.

Once we have such a statement, an obvious inductive argument implies that we have abelian $K^{O_s(1)}$-approximate groups $A_1, \dots, A_r \subset A^{O_s(1)}$ with $r \leq K^{O_s(1)}$ such that $A \subset A_1 \cdots A_r$. Theorem 4.1.2 then implies that there exist abelian progressions $P_1, \dots, P_r$ of rank at most $K^{O_s(1)}$ such that

$$A \subset P_1 \cdots P_r \subset A^{K^{O_s(1)}}. \tag{6.1.1}$$

Since $P_1 \cdots P_r$ can be viewed as an ordered progression, Proposition 5.6.4 then gives a nilprogression $P_{\text{nil}}(x; L)$ of rank at most $K^{O_s(1)}$ such that

$$P_1 \cdots P_r \subset P_{\text{nil}}(x; L) \subset \overline{P}(x; L) \subset (P_1 \cdots P_r)^{K^{O_s(1)}}, \tag{6.1.2}$$

and hence $A \subset P_{\text{nil}}(x; L) \subset \overline{P}(x; L) \subset A^{K^{O_s(1)}}$, as required by the torsion-free case of Theorem 6.1.1.

We spend the rest of this section sketching a proof of Proposition 6.1.3 (which also follows from the more general Proposition 6.6.1 below). To start with, rather than placing A inside a product of lower-step approximate groups, we find a product of lower-step approximate groups that is a large subset of a bounded power of A, as follows.

Proposition 6.1.4 *Let $K, s \geq 2$, let G be a torsion-free s-step nilpotent group, and suppose that $A \subset G$ is a finite K-approximate group. Then there exist $r \leq K^{O(1)}$ and finite $K^{O(1)}$-approximate groups*

$$A_0, \ldots, A_r \subset A^{O(1)},$$

each of which generates a subgroup of step less than s, such that

$$|A_0 \cdots A_r| \geq \exp(-K^{O(1)})|A|.$$

This is rather reminiscent of the proof of the Freiman–Green–Ruzsa theorem, Theorem 4.1.2. There, our aim was to prove that A was contained in a low-dimensional progression P. However, as a first step we proved Theorem 4.1.3, which showed instead that there was a large low-dimensional progression P contained in a bounded sumset of A. We then used Chang's covering argument (Proposition 4.7.1) to pass from the latter statement to the former. It turns out that a very similar argument allows us to pass from Proposition 6.1.4 to Proposition 6.1.3; we omit that argument from our sketch of Proposition 6.1.3, since it is essentially identical in the more general setting of Proposition 6.6.1.

Proposition 6.1.4 is an almost immediate consequence of the following two results.

Proposition 6.1.5 *Let $K, s \geq 2$, let G be a torsion-free s-step nilpotent group, and suppose that $A \subset G$ is a finite K-approximate group. Write $\pi : G \to G/[G,G]$ for the quotient homomorphism and, noting that $G/[G,G]$ is abelian and $\pi(A)$ is a K-approximate group, let $H \subset \pi(A^4)$ and $x_1, \ldots, x_r \in \pi(A^4)$ be the subgroup and elements given by applying Theorem 4.1.3 to $\pi(A)$. Then*

$$\left|(A^{16} \cap \pi^{-1}(H)) \prod_{i=1}^{r} (A^{22} \cap \pi^{-1}(\langle x_i \rangle))\right| \geq \frac{|A|}{\exp K^{O(1)}}.$$

We make no attempt to optimise the values of the exponents 16 and 22 in Proposition 6.1.5; all that matters to us is that they are absolute constants.

Lemma 6.1.6 *Let G be an s-step nilpotent group, and write $\pi : G \to G/[G,G]$ for the quotient homomorphism.*

(i) *For every $x \in G/[G,G]$ the subgroup $\pi^{-1}(\langle x \rangle)$ has step at most $s - 1$.*

(ii) If $H < G/[G,G]$ is a finite subgroup then $[\pi^{-1}(H),\dots,\pi^{-1}(H)]_s$ is generated by a union of finite subgroups. In particular, if G_s has no torsion then $\pi^{-1}(H)$ has step at most $s-1$.

Remark 6.1.7 The final conclusion of Lemma 6.1.6 (ii) does not necessarily hold if G_s has elements of finite order. For example, if G is a finite nilpotent group then $G/[G,G]$ is finite but $\pi^{-1}(G/[G,G]) = G$ does not have step lower than that of G.

Proof of Proposition 6.1.4 It follows from Proposition 2.6.5 that the sets $A^{16} \cap \pi^{-1}(H)$ and $A^{22} \cap \pi^{-1}(\langle x_i \rangle)$ appearing in Proposition 6.1.5 are $K^{O(1)}$-approximate groups. Proposition 6.1.4 therefore follows from Proposition 6.1.5 and Lemma 6.1.6. □

We prove Proposition 6.1.5 and Lemma 6.1.6 in the next section.

6.2 Details of the Torsion-Free Case

In this section we prove Proposition 6.1.5 and Lemma 6.1.6. In proving Proposition 6.1.5 we make use of the following observation, which is inspired by a lemma of Tao [62, Lemma 7.7].

Lemma 6.2.1 *Let G be a group, let $N \lhd G$ be a normal subgroup, and let $\pi : G \to G/N$ be the quotient homomorphism. Let A be a symmetric subset of G, and define a map $\varphi : \pi(A) \to A$ by choosing, for each element $x \in \pi(A)$, an element $\varphi(x) \in A$ such that $\pi(\varphi(x)) = x$. Then for every $a \in A$ we have*

$$a \in \left(A^2 \cap N\right) \varphi(\pi(a)), \tag{6.2.1}$$

and for every $x, y \in G/N$ with $x, y, xy \in \pi(A)$ we have

$$\varphi(xy) \in \varphi(x)\varphi(y)\left(A^3 \cap N\right). \tag{6.2.2}$$

Remark The condition (6.2.1) can be thought of as saying that φ is an inverse to π 'modulo $A^2 \cap N$', while (6.2.2) can be thought of as saying that φ is a homomorphism 'modulo $A^3 \cap N$'. We will apply Lemma 6.2.1 in the case where A is an approximate group, when both $A^2 \cap N$ and $A^3 \cap N$ will also be approximate groups by Proposition 2.6.5.

Proof Lemma 6.2.1 is essentially just an observation: by definition of φ we have $a\varphi(\pi(a))^{-1} \in A^2 \cap N$ and $\varphi(y)^{-1}\varphi(x)^{-1}\varphi(xy) \in A^3 \cap N$. □

Lemma 6.2.2 *Let G be a group, let $N \lhd G$ be a normal subgroup, and let $\pi : G \to G/N$ be the quotient homomorphism. Let A be a finite symmetric subset of G, and let $P \subset \pi(A^m)$. Suppose that $|P| \geq c|\pi(A)|$. Then $|\pi^{-1}(P) \cap A^{m+2}| \geq c|A|$.*

Proof Let $Y \subset A^m$ be such that π is injective on Y and $\pi(Y) = P$. Then we have $Y(A^2 \cap N) \subset \pi^{-1}(P) \cap A^{m+2}$ by definition, and

$$\begin{aligned} |Y(A^2 \cap N)| &= |P||A^2 \cap N| \\ &\geq c|\pi(A)||A^2 \cap N| \\ &\geq c|A| \qquad\qquad \text{(by Lemma 2.6.2)}, \end{aligned}$$

and so the lemma follows. □

Proof of Proposition 6.1.5 What Theorem 4.1.3 gives precisely is that there exists a finite subgroup $H \subset G/[G,G]$ and a progression $P = \{x_1^{\ell_1} \cdots x_r^{\ell_r} : |\ell_i| \leq L_i\}$ with $r \leq K^{O(1)}$ such that $HP \subset \pi(A^4)$ and $|HP| \geq \exp(-K^{O(1)})|\pi(A)|$. Lemma 6.2.2 therefore implies that

$$|\pi^{-1}(HP) \cap A^6| \geq \exp(-K^{O(1)})|A|. \tag{6.2.3}$$

We claim, in addition, that

$$\pi^{-1}(HP) \cap A^6 \subset \left(A^{16} \cap \pi^{-1}(H)\right) \prod_{i=1}^{r} \left(A^{22} \cap \pi^{-1}(\langle x_i \rangle)\right), \tag{6.2.4}$$

which combines with (6.2.3) to prove the proposition.

Define $\varphi : \pi(A^6) \to A^6$ by choosing, for each element $x \in \pi(A^6)$, an element $\varphi(x) \in A^6$ such that $\pi(\varphi(x)) = x$, taking care to choose $\varphi(x) \in A^4$ whenever $x \in \pi(A^4)$. Suppose that $a \in \pi^{-1}(HP) \cap A^6$, so that there exist $h \in H$ and $\ell_1, \ldots, \ell_r \in \mathbb{Z}$ such that $\pi(a) = hy_1^{\ell_1} \cdots y_r^{\ell_r}$. It follows from (6.2.1) that

$$a \in \left(A^{12} \cap [G,G]\right) \varphi(hx_1^{\ell_1} \cdots x_r^{\ell_r}),$$

and hence by repeated application of (6.2.2) that

$$\begin{aligned} a &\in \left(A^{12} \cap [G,G]\right) \varphi(h) \prod_{i=1}^{r} \varphi(x_i^{\ell_i}) \left(A^{18} \cap [G,G]\right) \\ &\subset \left(A^{16} \cap \pi^{-1}(H)\right) \prod_{i=1}^{r} \left(A^{22} \cap \pi^{-1}(\langle x_i \rangle)\right). \end{aligned}$$

This gives (6.2.4), and hence the proposition, as claimed. □

Proof of Lemma 6.1.6

(i) Pick $y \in G$ such that $\pi(y) = x$, and note that $\pi^{-1}(\langle x \rangle) = \langle y \rangle [G, G]$. Given $z_1, \ldots, z_s \in \pi^{-1}(\langle x \rangle)$, there therefore exist $n_1, \ldots, n_s \in \mathbb{Z}$ and $c_1, \ldots, c_d \in [G, G]$ such that each $z_i = y^{n_1} c_1$. By Lemma 5.5.2 and Corollary 5.2.9, this implies that

$$\begin{aligned}[\, z_1, \ldots, z_s \,] &= [\, y^{n_1} c_1, \ldots, y^{n_s} c_s \,] \\ &= [\, y, \ldots, y \,]_s^{n_1 \cdots n_s} \\ &= 1,\end{aligned}$$

and so $\pi^{-1}(\langle x \rangle)$ has step at most $s - 1$, as required.

(ii) By definition the group $[\, \pi^{-1}(H), \ldots, \pi^{-1}(H) \,]_s$ is generated by

$$X = \{[\, g_1 \ldots, g_s \,] : g_i \in \pi^{-1}(H)\}.$$

Lemma 5.5.2 and Corollary 5.2.9 imply that given $g_2, \ldots, g_s \in \pi^{-1}(H)$ there exists a homomorphism $\alpha : H \to G_s$ such that $[g_1, g_2, \ldots, g_s] = \alpha(\pi(g_1))$ for every $g_1 \in \pi^{-1}(H)$, and in particular that

$$\{[\, g_1, \ldots, g_s \,] : g_1 \in \pi^{-1}(H)\} = \alpha(H).$$

Since H is a finite subgroup, so is $\alpha(H)$, and so X is a union of finite subgroups and the lemma is proved.

□

Remark The proof of the first part of Lemma 6.1.6 shows more generally that if N is a subgroup of G of step less than s then $N[G, G]$ also has step less than s.

6.3 Abelian p-Groups

We now move on to consider nilpotent groups that contain elements of finite order. Unfortunately, Proposition 6.1.3 does not necessarily hold in this setting, as illustrated by the following example.

Example 6.3.1 Let F be the free product of n copies of the cyclic group with two elements, and let G be the quotient $F/[F, [F, F]]$. Then G is a finite 2-step nilpotent group, and in particular a K-approximate subgroup of a 2-step nilpotent group for every $K \geq 1$. However, G cannot be expressed as a product of $O(1)$ abelian sets as $n \to \infty$.

The specific part of the proof of Proposition 6.1.3 that breaks down is where we apply Lemma 6.1.6 (ii), since, as we note in Remark 6.1.7, $[\pi^{-1}(H),\ldots,\pi^{-1}(H)]_s$ need not be trivial if G has elements of finite order. What Lemma 6.1.6 (ii) does tell us is that $[\pi^{-1}(H),\ldots,\pi^{-1}(H)]_s$ is generated by a union of finite subgroups. In this section we investigate how to control such groups.

We start by restricting attention to *p-groups*, where the details are cleanest. Given a prime p, a *p-group* is a group in which every element has order a power of p. Such groups are a natural archetypal setting in which to study nilpotent groups with elements of finite order, since p-groups are always nilpotent and, moreover, every finite nilpotent group is a direct product of p-groups. Since we do not actually need this fact for our arguments we omit the proof, referring the reader instead to [38, Theorem 10.3.4], for example.

The main result of this section is the following.

Proposition 6.3.2 *Let Γ be an abelian p-group of rank r and suppose that $X \subset \Gamma$ is a union of subgroups of Γ. Then $\langle X\rangle \subset rX$.*

We note in Remark 6.3.8 that the assumption in Proposition 6.3.2 that Γ is a p-group is necessary.

To illustrate the utility of Proposition 6.3.2, before proving it we use it to sketch a proof of the following variant of Proposition 6.1.4 for p-groups, which we do not use in our subsequent arguments but which gives an idea of how we can adapt the induction step from the torsion-free case of Theorem 6.1.1.

Proposition 6.3.3 *Let $K, s \geq 2$, let G be an s-step nilpotent p-group, and suppose that $A \subset G$ is a finite K-approximate group. Then there exist a subgroup $N \subset A^{K^{e^{O(s)}}}$ normalised by A, a natural number $r \leq K^{O(1)}$, and finite $K^{O(1)}$-approximate groups $A_0,\ldots,A_r \subset A^{O(1)}$ such that $[A_i,\ldots,A_i]_s \subset N$ for each i, and such that*

$$|A_0\cdots A_r| \geq \exp(-K^{O(1)})|A|.$$

The difference between the conclusions of Propositions 6.1.4 and 6.3.3 is that, whereas in Proposition 6.1.4 the sets A_i all generate subgroups of step less than s, in Proposition 6.3.3 they only generate subgroups of step less than s modulo N. Nonetheless, the fact that $N \subset A^{K^{e^{O(s)}}}$ means that we at least have some control over this error N, giving some hope of being able to induct on s in the proof of Theorem 6.1.1 for p-

groups. This does indeed turn out to be possible, as we shall see in the next few sections.

Sketch proof of Proposition 6.3.3 We define $r \leq K^{O(1)}$ and the sets $A_0, \ldots, A_r \subset A^{O(1)}$ in exactly the same way as in the proof of Proposition 6.1.4. As in that proof, the sets $A_1, \ldots, A_r$ automatically generate subgroups of step less than s. However, all we can say about A_0 is that, if we write $\pi : G \to G/[G,G]$ for the quotient homomorphism, there is a subgroup $H \subset \pi(A^4)$ such that $A_0 \subset \pi^{-1}(H)$.

Let $Y \subset A^4$ be such that $\pi(Y) = H$. It follows from Lemma 5.5.2 that $[\,\pi^{-1}(H), \ldots, \pi^{-1}(H)\,]_s$ is generated by the set

$$X = \{[\,y_1, \ldots, y_s\,] : y_i \in Y\},$$

and that for every fixed $y_2, \ldots, y_s \in Y$ there exists a homomorphism $\alpha : H \to G_s$ such that

$$\{[\,y_1, \ldots, y_s\,] : y_1 \in Y\} = \alpha(H).$$

This implies that X is a union of subgroups. Now Lemma 5.3.6 implies that

$$X \subset A^{2^{s+1}s} \cap G_s. \tag{6.3.1}$$

Since G is s-step nilpotent, G_s is abelian, and even central in G, by Proposition 5.2.4 and Corollary 5.2.9. Moreover, the set $A^{2^{s+1}s} \cap G_s$ is a $K^{2^{s+2}s}$-approximate group by Proposition 2.6.5. Theorem 4.1.2 and (6.3.1) therefore imply that there is a subgroup

$$H \subset A^{2^{s+3}s} \cap G_s \tag{6.3.2}$$

and a progression $P \subset G_s$ of rank at most $K^{O(2^s s)}$ such that $X \subset HP$. The bound on the rank of P implies that $\langle HP \rangle / H$ is an abelian p-group of rank at most $K^{O(2^s s)}$. Applying Proposition 6.3.2 modulo H therefore gives

$$\begin{aligned}
[\,\pi^{-1}(H), \ldots, \pi^{-1}(H)\,]_s &= \langle X \rangle \\
&\subset X^{K^{O(2^s s)}} H && \text{(by Proposition 6.3.2)} \\
&\subset A^{K^{O(2^s s)} + 2^{s+3}s} && \text{(by (6.3.1) and (6.3.2))} \\
&\subset A^{K^{e^{O(s)}}} && \text{(since } K, s \geq 2\text{).}
\end{aligned}$$

Since $[\,\pi^{-1}(H), \ldots, \pi^{-1}(H)\,]_s$ is central in G it is in particular normal, so we may take $N = [\,\pi^{-1}(H), \ldots, \pi^{-1}(H)\,]_s$. □

In the remainder of this section we prove Proposition 6.3.2. The proof rests on the following properties of finite abelian p-groups. We define a *maximal subgroup* of a group G to be a subgroup M such that the only subgroup $H < G$ satisfying $M \subsetneq H \subset G$ is G itself.

Lemma 6.3.4 *Let Γ be a finite abelian p-group. Then a subgroup $H \subsetneq \Gamma$ is maximal if and only if $\Gamma/H \cong \mathbb{Z}/p\mathbb{Z}$.*

Proof The fundamental theorem of finitely generated abelian groups (Theorem 1.5.1) implies that

$$\Gamma/H \cong \bigoplus_{i=1}^{k} \mathbb{Z}/p^{n_i}\mathbb{Z}$$

for some $k \in \mathbb{N}_0$ and $n_1, \ldots, n_k \in \mathbb{N}$, from which the lemma is immediate. □

Proposition 6.3.5 *Let Γ be a finite abelian p-group. Then a subset $S \subset \Gamma$ generates Γ if and only if $S + (p \cdot \Gamma)$ generates Γ.*

Proof If S generates Γ then of course $S+(p\cdot\Gamma)$ does as well. Conversely, if S does not generate Γ then S is contained in some maximal subgroup H of Γ. Since $\Gamma/H \cong \mathbb{Z}/p\mathbb{Z}$ by Lemma 6.3.4, this means that $p \cdot \Gamma \subset H$, and so $S + (p\cdot\Gamma)$ generates a subgroup of H, and hence does not generate Γ. □

Corollary 6.3.6 *Let Γ be a finite abelian p-group. Then all minimal generating sets for Γ have the same size.*

Proof Since every non-zero element in $\Gamma/(p \cdot \Gamma)$ has order p, the fundamental theorem of finitely generated abelian groups (Theorem 1.5.1) implies that $\Gamma/(p \cdot \Gamma) \cong \mathbb{F}_p^n$ for some $n \in \mathbb{N}$. Proposition 6.3.5 then implies that a set $S \subset \Gamma$ generates Γ if and only if its image modulo $p \cdot \Gamma$ generates $\mathbb{F}_p^n$. It follows that S is a minimal generating set for Γ if and only if each element of S lies in a different coset of $p\cdot\Gamma$ and the image of S modulo $p \cdot \Gamma$ is a basis for $\mathbb{F}_p^n$. In particular, every minimal generating set for Γ has size n. □

Lemma 6.3.7 *If Γ is an abelian p-group and Γ' is a subgroup of Γ then the rank of Γ' is at most the rank of Γ.*

Proof Let $x_1, \ldots, x_r$ be a generating set of minimum size for Γ. We will show by induction on r that the rank of Γ' is at most r, this being trivial when $r = 0$.

Let m be the largest integer with the property that $\Gamma' < p^m \cdot \Gamma$. The subgroup $p^m \cdot \Gamma$ is generated by the set $p^m x_1, \dots, p^m x_r$, and so has rank at most r. Therefore, upon replacing Γ by $p^m \cdot \Gamma$ if necessary, we may assume that Γ' is not contained in $p \cdot \Gamma$, and hence that there is some $y \in \Gamma'$ such that $px \neq y$ for every $x \in \Gamma$.

Since $x_1, \dots, x_r$ generate Γ we may write $y = \ell_1 x_1 + \cdots + \ell_r x_r$, and by definition of y there must be at least one i for which ℓ_i is not divisible by p. Without loss of generality we shall assume that ℓ_r is not divisible by p. Since the order of x_r is a power of p, the integer ℓ_r has a multiplicative inverse, say q, modulo the order of the x_r. This implies that

$$x_r = q(y - \ell_1 x_1 - \cdots - \ell_{r-1} x_{r-1}),$$

and in particular that the quotient $\Gamma / \langle y \rangle$ is generated by

$$x_1 + \langle y \rangle, \dots, x_{r-1} + \langle y \rangle,$$

and hence has rank at most $r - 1$. The group $\Gamma' / \langle y \rangle$ is isomorphic to a subgroup of $\Gamma / \langle y \rangle$, and hence has rank at most $r - 1$ by induction. It follows that Γ' has rank at most r, as claimed. □

We invite the reader to prove a version of Lemma 6.3.7 for arbitrary finite p-groups in Exercise 6.2.

Proof of Proposition 6.3.2 Lemma 6.3.7 implies that $\langle X \rangle$ is of rank at most r, and so Corollary 6.3.6 implies that there exist elements $x_1, \dots, x_r \in X$ that generate $\langle X \rangle$. It follows that $\langle X \rangle = \langle x_1 \rangle + \cdots + \langle x_r \rangle$. However, the assumption that X is a union of subgroups implies that $\langle x_i \rangle \subset X$ for each i, and so the lemma is proved. □

Remark 6.3.8 The assumption in Proposition 6.3.2 that Γ is a p-group is necessary. The statement fails, for example, if $\Gamma = \mathbb{Z}/6\mathbb{Z}$ and X is the union of the subgroups $\{0, 2, 4\}$ and $\{0, 3\}$. This is due to the failure of Corollary 6.3.6 in this setting (see Exercise 6.2).

6.4 Multi-Variable Homomorphisms into Abelian Groups

The purpose of this section is to modify the results of the previous section to work in nilpotent groups that are not necessarily p-groups. Unfortunately, as we observed in Remark 6.3.8, Proposition 6.3.2 fails when Γ is not a p-group, which causes problems when trying to place $[\pi^{-1}(H),\ldots,\pi^{-1}(H)]_s$ inside $A^{O_{K,s}(1)}$ in results such as Proposition 6.3.3.

Nonetheless, it turns out that we can still use the fact that the subgroup $[\pi^{-1}(H),\ldots,\pi^{-1}(H)]_s$ is generated by images of H under multi-variable homomorphisms, thanks to the following result.

Proposition 6.4.1 *Let $k > 0$ be an integer, let $U_1,\ldots,U_k$ be finite groups and let Γ be an abelian group of rank at most r. Let $\varphi : U_1 \times \cdots \times U_k \to \Gamma$ be a map that is a homomorphism in each variable. Then*

$$\langle \varphi(U_1,\ldots,U_k)\rangle \subset r\,\varphi(U_1,\ldots,U_k).$$

When Γ is a p-group, Proposition 6.4.1 follows from exactly the same arguments we used to prove Proposition 6.3.3. As we are about to see, in general we can reduce to that case using the fundamental theorem of finitely generated abelian groups (Theorem 1.5.1).

Lemma 6.4.2 *For $i = 1,\ldots,k$ let $U_i = \bigoplus_p U_{ip}$ be a direct sum of abelian p-groups U_{ip}, and let Γ be a finitely generated abelian group, expressed using the fundamental theorem of finitely generated abelian groups as the direct sum $\Gamma = \Gamma_0 \oplus \bigoplus_p \Gamma_p$ of a torsion-free abelian group Γ_0 and abelian p-groups Γ_p. Let $\varphi : U_1 \times \cdots \times U_k \to \Gamma$ be a map that is a homomorphism in each variable.*

Given $(u_1,\ldots,u_k) \in U_1 \times \cdots \times U_k$, for each i write $u_i = \sum_p u_{ip}$ with $u_{ip} \in U_{ip}$ for every prime p. Then we have

$$\varphi(u_{1p},\ldots,u_{kp}) \in \Gamma_p \tag{6.4.1}$$

for every prime p, and we have

$$\varphi(u_1,\ldots,u_k) = \sum_p \varphi(u_{ip},\ldots,u_{kp}). \tag{6.4.2}$$

Proof The fact that φ is a homomorphism in each variable implies that

$$\varphi(u_1,\ldots,u_k) = \sum_{p_1,\ldots,p_k} \varphi(u_{1p_1},\ldots,u_{kp_k}).$$

It also implies that the order of $\varphi(u_{1p_1},\ldots,u_{kp_k})$ divides the order of each u_{ip_i}. This implies that for every fixed p the order of $\varphi(u_{1p},\ldots,u_{kp})$ is a power of p, and so (6.4.1) holds. It also implies that if the p_i are not all equal then the order of $\varphi(u_{1p_1},\ldots,u_{kp_k})$ is 1, giving (6.4.2). □

Proposition 6.4.3 *Let Γ be a finite abelian group of rank r expressed as a direct sum $\Gamma = \bigoplus_p \Gamma_p$ of p-groups. For each p, suppose that $X_p \subset \Gamma_p$ is a union of subgroups of Γ_p, and write $X = \bigoplus_p X_p$. Then $\langle X \rangle \subset rX$.*

Proof Since $\Gamma_p \cong \Gamma / \bigoplus_{q \neq p} \Gamma_q$, the rank of each Γ_p is at most r, and so Proposition 6.3.2 implies that $\langle X_p \rangle \subset rX_p$, from which it easily follows that $\langle X \rangle \subset rX$. □

Proof of Proposition 6.4.1 Since the range of φ is abelian, each homomorphism $U_i \to \Gamma$ obtained by allowing the i variable to vary and fixing the other variables is trivial on $[U_i, U_i]$. In particular, φ factors through the quotients $U_i/[U_i, U_i]$, and so there exists a map

$$\varphi' : U_1/[U_1, U_1] \times \cdots \times U_k/[U_k, U_k] \to \Gamma$$

such that

$$\varphi'(U_1/[U_1, U_1], \ldots, U_k/[U_k, U_k]) = \varphi(U_1, \ldots, U_k).$$

We may therefore assume that each U_i is abelian, and hence, by Theorem 1.5.1, of the form $U_i = \bigoplus_p U_{ip}$ with each U_{ip} a p-group. Lemma 6.4.2 therefore implies that, writing $\Gamma = \Gamma_0 \oplus \bigoplus_p \Gamma_p$ with Γ_0 having no torsion and each Γ_p a p-group (again by Theorem 1.5.1), we have

$$\varphi(U_1, \ldots, U_k) = \bigoplus_p \varphi(U_{ip}, \ldots, U_{kp})$$

with $\varphi(U_{ip}, \ldots, U_{kp}) \subset \Gamma_p$ for every prime p. Since $\bigoplus_p \Gamma_p \cong \Gamma/\Gamma_0$, the group $\bigoplus_p \Gamma_p$ has rank at most r, and so it follows from Proposition 6.4.3 that $\langle \varphi(U_1, \ldots, U_k) \rangle \subset r\,\varphi(U_1, \ldots, U_k)$, as required. □

6.5 Placing Arbitrary Subgroups inside Normal Subgroups

In Proposition 6.4.1 we showed that certain images of multi-variable homomorphisms in abelian groups can be placed efficiently inside subgroups. However, our ultimate aim is to prove Theorem 6.1.1, where the only subgroup we are permitted in the conclusion is a *normal* subgroup. In order for Proposition 6.4.1 to be of any use, therefore, we must show that the subgroup of its conclusion can be placed efficiently inside a subgroup that is normal in the ambient group. We achieve this in this section via the following proposition.

Proposition 6.5.1 *Let G be an s-step nilpotent group generated by a finite K-approximate group A. Let $H_1, \dots, H_r$ be finite groups and suppose that $\alpha : H_1 \times \cdots \times H_r \to G$ is a homomorphism in each variable with the property that*

$$\alpha(H_1, \dots, H_r) \subset A^m. \tag{6.5.1}$$

Then there exists a normal subgroup $N \lhd G$ with the property that

$$\alpha(H_1, \dots, H_r) \subset N \subset A^{K^{e^{O(s)}} m}.$$

It is worth recording that this implies in particular that an arbitrary subgroup can be placed efficiently inside a normal subgroup, as follows.

Corollary 6.5.2 *Let G be an s-step nilpotent group generated by a finite K-approximate group A. Let $H \subset A^m$ be a subgroup of G. Then there exists a normal subgroup $N \lhd G$ such that $H \subset N \subset A^{K^{e^{O(s)}} m}$.*

Proof Apply Proposition 6.5.1 in the case $r = 1$ with α the inclusion homomorphism $\alpha : H \hookrightarrow G$. □

The proof of Proposition 6.5.1 rests on yet another lemma about multilinearity of commutators, as follows.

Lemma 6.5.3 *Let G be a group. Then the map*

$$\begin{array}{ccc} Z_n(G) \times G \times \cdots \times G & \to & Z_1(G) \\ (z, g_1, \dots, g_{n-1}) & \mapsto & [z, g_1, \dots, g_{n-1}] \end{array}$$

is a homomorphism in each variable.

Proof We proceed by induction on n, noting that the case $n = 1$ is trivial. We may therefore assume that the map

$$\begin{array}{ccc} Z_n(G) \times G \times \cdots \times G & \to & Z_2(G) \\ (z, g_1, \dots, g_{n-2}) & \mapsto & [z, g_1, \dots, g_{n-2}] \end{array} \tag{6.5.2}$$

is a homomorphism in each variable modulo $Z_1(G)$.

Lemma 5.5.3 implies that if A and B are subgroups of G whose commutator $[A, B]$ lies in the centre of G then the commutator map $A \times B \to [A, B]$ is a homomorphism in each variable. This implies in particular that the map

$$\begin{array}{ccc} [Z_n(G), G, \dots, G]_{n-1} \times G & \to & Z_1(G) \\ ([z, g_1, \dots, g_{n-2}], g_{n-1}) & \mapsto & [z, g_1, \dots, g_{n-1}] \end{array} \tag{6.5.3}$$

is a homomorphism in each variable, with $Z_1(G)$ lying in the kernel of each of these homomorphisms. The result therefore follows from the fact that the map (6.5.2) is a homomorphism in each variable modulo $Z_1(G)$. □

Proof of Proposition 6.5.1 If $K < 2$ then $A = G$ and the proposition is trivial, so we may assume that $K \geq 2$. Let $n \in \mathbb{N}$ be minimal such that

$$\alpha(H_1, \ldots, H_r) \subset Z_n(G).$$

We actually prove that there exists a normal subgroup $N \lhd G$ with the property that

$$\alpha(H_1, \ldots, H_r) \subset N \subset A^{\sum_{i=1}^{n} K^{e^{O(i)} m}}. \tag{6.5.4}$$

Since $n \leq s$ by Corollary 5.2.9, this implies in particular that

$$N \subset A^{sK^{e^{O(s)} m}} \subset A^{K^{e^{O(s)} m}},$$

as required by the proposition.

The case $n = 0$ being trivial, we may assume that $n \geq 1$ and proceed by induction. It follows from Lemma 6.5.3 that the map

$$\begin{array}{rccl} \varphi : & H_1 \times \cdots \times H_r \times G \times \cdots \times G & \to & Z_1(G) \\ & (h_1, \ldots, h_r, g_1, \ldots, g_{n-1}) & \mapsto & [\alpha(h_1, \ldots, h_r), g_1, \ldots, g_{n-1}] \end{array}$$

is a homomorphism in each variable. Moreover, since the image of φ is abelian, if some $g_i \in [G, G]$ then $\varphi(h_1, \ldots, h_r, g_1, \ldots, g_{n-1}) = 1$. There therefore exists a map

$$\psi : H_1 \times \cdots \times H_r \times G/[G, G] \times \cdots \times G/[G, G] \to Z_1(G)$$

that is a homomorphism in each variable such that, writing π for the quotient $\pi : G \to G/[G, G]$, we have

$$[\alpha(h_1, \ldots, h_r), g_1, \ldots, g_{n-1}] = \psi(h_1, \ldots, h_r, \pi(g_1), \ldots, \pi(g_{n-1})).$$

We claim that there exists a central subgroup N_0 of G satisfying $N_0 \subset A^{K^{e^{O(n)m}}}$ such that

$$[\alpha(H_1, \ldots, H_r), G, \ldots, G]_n \subset N_0. \tag{6.5.5}$$

In order to prove this claim it suffices to exhibit a generating set Y for $G/[G, G]$ and a central subgroup $N_0 \subset A^{K^{e^{O(n)m}}}$ such that

$$\psi(H_1, \ldots, H_r, Y, \ldots, Y) \subset N_0. \tag{6.5.6}$$

Of course, $\pi(A)$ generates $G/[G, G]$. Moreover, $\pi(A)$ is a K-approximate group, and so Theorem 4.1.2 implies that there exist an integer $k \leq$

$K^{O(1)}$, a subgroup $H \subset \pi(A^4)$ and elements $x_1, \ldots, x_k \in \pi(A^4)$ such that $H \cup \{x_1, \ldots, x_k\}$ generates $G/[G,G]$. We will prove (6.5.6) with $Y = H \cup \{x_1, \ldots, x_k\}$.

Writing $U_0 = H$ and $U_i = \{x_i\}$ for $i = 1, \ldots, k$, we have

$$\psi(H_1, \ldots, H_r, U_{i_1}, \ldots, U_{i_{n-1}}) \subset [\,\alpha(H_1, \ldots, H_r), A^4, \ldots, A^4\,]_n$$

for all possible choices of i_j. However, the definition of n implies that $[\,\alpha(H_1, \ldots, H_r), G, \ldots, G\,]_n \subset Z(G)$, and so (6.5.1) and Lemma 5.3.6 imply that

$$[\,\alpha(H_1, \ldots, H_r), A^4, \ldots, A^4\,]_n \subset A^{2^{n-1}(m+4n)} \cap Z(G).$$

Bounding $2^{n-1}(m+4n)$ by $4^n m$ for notational convenience, this implies that

$$\psi(H_1, \ldots, H_r, U_{i_1}, \ldots, U_{i_{n-1}}) \subset A^{4^n m} \cap Z(G) \qquad (6.5.7)$$

for all choices of i_j. Proposition 2.6.5 implies that $A^{4^n m} \cap Z(G)$ is a $K^{e^{O(n)}m}$-approximate group. Theorem 4.1.2 therefore implies that $A^{4^n m} \cap Z(G)$ is contained inside an abelian coset progression ZQ, with Z a subgroup satisfying $Z \subset A^{4^{n+1}m} \cap Z(G)$ and Q a progression of rank at most $K^{e^{O(n)}m}$, so that (6.5.7) implies that

$$\psi(H_1, \ldots, H_r, U_{i_1}, \ldots, U_{i_{n-1}}) \subset A^{4^n m} \cap ZQ.$$

Since Z is central in G it is certainly normal in G, and so we may apply Proposition 6.4.1 in the quotient G/Z. Since each U_{i_j} is either a singleton or a finite group, the bound on the rank of Q means that this gives

$$\begin{aligned} &\langle \psi(H_1, \ldots, H_r, U_{i_1}, \ldots, U_{i_{n-1}}) \rangle \\ &\qquad \subset \psi(H_1, \ldots, H_r, U_{i_1}, \ldots, U_{i_{n-1}})^{K^{e^{O(n)}m}} \cdot Z \\ &\qquad \subset A^{K^{e^{O(n)}m}}. \end{aligned}$$

Taking N_0 to be the product of the central subgroups Z and

$$\langle \psi(H_1, \ldots, H_r, U_{i_1}, \ldots, U_{i_{n-1}}) \rangle$$

as the indices i_j vary over all possibilities, we therefore have (6.5.6) and

$$N_0 \subset A^{K^{e^{O(n)}m}}. \qquad (6.5.8)$$

Now by (6.5.5) the image of $\alpha(H_1, \ldots, H_r)$ in G/N_0 lies in $Z_{n-1}(G/N_0)$, and so by induction we may assume that there is some normal subgroup $N \lhd G$ containing N_0 such that

$$\alpha(H_1, \ldots, H_r) \subset N \subset A^{\sum_{i=1}^{n-1} K^{e^{O(i)}m}} \cdot N_0.$$

Combined with (6.5.8), this proves (6.5.4), as claimed, and the proposition follows. □

6.6 Conclusion of the General Case

In this section we prove Theorem 6.1.1. The key ingredient is the following generalisation of Proposition 6.1.3.

Proposition 6.6.1 *Let $m > 0$ and $s \geq \tilde{s} \geq 2$ be integers, and let $K, \tilde{K} \geq 2$. Let G be an s-step nilpotent group generated by a finite K-approximate group A, and let $\tilde{A} \subset A^m$ be a $\tilde{K}$-approximate group that generates an $\tilde{s}$-step nilpotent subgroup $\tilde{G}$ of G. Then there exist an integer $k \leq \tilde{K}^{O(1)}$, a normal subgroup $N \lhd G$ satisfying $N \subset A^{K^{e^{O(s)}m}}$, and $\tilde{K}^{O(1)}$-approximate groups $A_1, \dots, A_k \subset \tilde{A}^{O(1)}$ such that $[\,A_i, \dots, A_i\,]_{\tilde{s}} \subset N$ for all i, and such that $\tilde{A} \subset A_1 \cdots A_k$.*

Note that if the group G in Proposition 6.6.1 has no torsion then the finite subgroup N is automatically trivial, and so we recover Proposition 6.1.3.

In proving Proposition 6.6.1 we use Lemma 6.1.6 in the same way as in the proof of Proposition 6.1.3, but supplemented by the following result.

Proposition 6.6.2 *Let $m > 0$ and $s \geq \tilde{s} \geq 2$ be integers, and let $K \geq 2$. Let G be an s-step nilpotent group generated by a finite K-approximate group A, and let $\tilde{G}$ be an $\tilde{s}$-step nilpotent subgroup of G. Write $\pi : \tilde{G} \to \tilde{G}/[\tilde{G}, \tilde{G}]$ for the quotient homomorphism, and suppose that $H \subset \pi(A^m \cap \tilde{G})$ is a finite group. Then there is a normal subgroup $N \lhd G$ such that $[\,\pi^{-1}(H), \dots, \pi^{-1}(H)\,]_{\tilde{s}} \subset N \subset A^{K^{e^{O(s)}m}}$.*

Proof It follows from Lemma 5.5.2 that the commutator form

$$[\ ,\dots,\ ]_{\tilde{s}} : \tilde{G} \times \cdots \times \tilde{G} \to \tilde{G}_{\tilde{s}}$$

is a homomorphism in each variable, and the fact that $\tilde{G}_{\tilde{s}}$ is abelian implies that it is equal to the identity whenever any of its components lies in $[\tilde{G}, \tilde{G}]$. There is therefore a map

$$\alpha : \frac{\tilde{G}}{[\tilde{G}, \tilde{G}]} \times \cdots \times \frac{\tilde{G}}{[\tilde{G}, \tilde{G}]} \to \tilde{G}_{\tilde{s}}$$

that is a homomorphism in each variable such that

$$[g_1, \ldots, g_{\tilde{s}}] = \alpha(g_1[\tilde{G}, \tilde{G}], \ldots, g_{\tilde{s}}[\tilde{G}, \tilde{G}])$$

for every $g_1, \ldots, g_{\tilde{s}} \in \tilde{G}$. In particular,

$$[\pi^{-1}(H), \ldots, \pi^{-1}(H)]_{\tilde{s}} = \langle \alpha(H, \ldots, H) \rangle$$

and

$$\begin{aligned} \alpha(H, \ldots, H) &\subset [A^m, \ldots, A^m]_{\tilde{s}} \\ &\subset A^{e^{O(s)}m} \qquad \text{(by Lemma 5.3.6).} \end{aligned}$$

The desired result therefore follows from Proposition 6.5.1. □

Proof of Proposition 6.6.1 Proposition 6.1.5 implies that there exist an integer $r \le \tilde{K}^{O(1)}$, elements $x_1, \ldots, x_r \in \tilde{A}^4$, and a set $X_0 \subset \tilde{A}^4$ such that $X_0[\tilde{G}, \tilde{G}]$ is a subgroup of G and such that

$$\left| \left(\tilde{A}^{16} \cap X_0[\tilde{G}, \tilde{G}]\right) \prod_{i=1}^{r} \left(\tilde{A}^{22} \cap \langle x_i \rangle [\tilde{G}, \tilde{G}]\right) \right| \ge \frac{|\tilde{A}|}{\exp \tilde{K}^{O(1)}}.$$

Write $B_0 = \tilde{A}^{16} \cap X_0[\tilde{G}, \tilde{G}]$ and $B_i = \tilde{A}^{22} \cap \langle x_i \rangle [\tilde{G}, \tilde{G}]$ for $i = 1, \ldots, r$, and note that each set B_i is a $\tilde{K}^{O(1)}$-approximate group by Proposition 2.6.5.

Chang's covering lemma (Proposition 4.7.1) implies that there exist $t \le \tilde{K}^{O(1)}$ and sets $S_1, \ldots, S_t \subset \tilde{A}$ with $|S_i| \le 2\tilde{K}$ such that

$$\tilde{A} \subset S_t \cdots S_1 B_0 \cdots B_r B_r \cdots B_0 S_1^{-1} \cdots S_{t-1}^{-1}.$$

Enumerating the elements of $\bigcup_i S_i$ as $u_1, \ldots, u_\ell$ in such a way that if $i < j$ then the elements of S_i appear before the elements of S_j, and writing $U_i = \{u_i^{-1}, 1, u_i\} \subset \tilde{A}$, we then have $\ell \le \tilde{K}^{O(1)}$ and

$$\tilde{A} \subset U_\ell \cdots U_1 B_0 \cdots B_r B_r \cdots B_0 U_1 \cdots U_\ell.$$

Each of the sets U_i is a 2-approximate group generating an abelian subgroup, whilst the sets $B_1, \ldots, B_r$ generate subgroups of step strictly less than $\tilde{s}$ by Lemma 6.1.6. Moreover, Proposition 6.6.2 implies that there is a normal subgroup $N \lhd G$ such that $[B_0, \ldots, B_0]_{\tilde{s}} \subset N \subset A^{K^{e^{O(s)}m}}$. The proposition therefore follows from taking $k = 2r + 2\ell$ and relabelling the sets U_i and B_j as $A_1, \ldots, A_k$. □

It is a fairly straightforward matter to apply Proposition 6.6.1 repeatedly and end up with the following result.

Proposition 6.6.3 *Let $m > 0$ and $s \geq \tilde{s} \geq 1$ be integers, and let $K, \tilde{K} \geq 2$. Let G be an s-step nilpotent group generated by a finite K-approximate group A, and let $\tilde{A} \subset A^m$ be a $\tilde{K}$-approximate group that generates an $\tilde{s}$-step nilpotent subgroup $\tilde{G}$ of G. Then there exist an integer $r \leq \tilde{K}^{e^{O(\tilde{s})}}$, a normal subgroup $N \lhd G$ satisfying $N \subset A^{\tilde{K}^{e^{O(\tilde{s})}} K^{e^{O(s)}} m}$, and $\tilde{K}^{e^{O(\tilde{s})}}$-approximate groups $A_1, \ldots, A_r \subset \tilde{A}^{e^{O(\tilde{s})}} N$ such that $[A_i, A_i] \subset N$ for all i, and such that $\tilde{A} \subset A_1 \cdots A_r N$.*

Proof If $\tilde{A}$ is abelian then the proposition is trivially true with $r = 1$, $A_1 = \tilde{A}$ and $N = \{1\}$. We may therefore assume that $s \geq \tilde{s} \geq 2$ and, by induction, that the proposition holds for all smaller values of $\tilde{s}$.

We start by rewriting the statement we are trying to prove as follows: there exist an integer $r \leq \tilde{K}^{\tilde{s} e^{O(\tilde{s})}}$, a normal subgroup $N \lhd G$ with $N \subset A^{\tilde{s} \tilde{K}^{e^{O(2\tilde{s})}} K^{e^{O(s)+O(\tilde{s})}} m}$, and $\tilde{K}^{e^{O(\tilde{s})}}$-approximate groups

$$A_1, \ldots, A_r \subset \tilde{A}^{e^{O(\tilde{s})}} N$$

such that $[A_i, A_i] \subset N$ for all i, and such that $\tilde{A} \subset A_1 \cdots A_r N$. This is exactly equivalent to the conclusion of the proposition, but writing the bounds in this way makes it easier to keep track of them through the induction. For the same reason, at various points in the argument we use the trivial observation that any quantity bounded by $O(1)$ is also bounded by $e^{O(1)}$.

Applying Proposition 6.6.1, we obtain an integer $r_0 \leq \tilde{K}^{O(1)} \leq \tilde{K}^{e^{O(1)}}$, a normal subgroup $N_0 \lhd G$ satisfying $N_0 \subset A^{K^{e^{O(s)}} m}$, and $\tilde{K}^{e^{O(1)}}$-approximate groups $\tilde{A}_1, \ldots, \tilde{A}_{r_0} \subset \tilde{A}^{O(1)} \subset \tilde{A}^{e^{O(1)}}$ such that

$$[\tilde{A}_i, \ldots, \tilde{A}_i]_{\tilde{s}} \subset N_0$$

for all i, and such that

$$\tilde{A} \subset \tilde{A}_1 \cdots \tilde{A}_{r_0}. \tag{6.6.1}$$

Writing $\rho : G \to G/N_0$ for the quotient homomorphism, it follows that the sets $\rho(\tilde{A}_1), \ldots, \rho(\tilde{A}_{r_0}) \subset \rho(\tilde{A})^{e^{O(1)}} \subset \rho(A)^{e^{O(1)} m}$ are of step at most $\tilde{s} - 1$. Since G/N_0 is generated by $\rho(A)$, the induction hypothesis therefore gives, for each $i = 1, \ldots, r_0$, an integer $r_i \leq \tilde{K}^{(\tilde{s}-1) e^{O(\tilde{s})}}$, a normal subgroup $N_i \lhd G$ containing N_0 and satisfying

$$\begin{aligned} N_i &\subset A^{(\tilde{s}-1)(\tilde{K}^{e^{O(1)}})^{e^{O(2\tilde{s}-2)}} K^{e^{O(s)+O(\tilde{s}-1)}(e^{O(1)} m)}} \cdot N_0 \\ &\subset A^{(\tilde{s}-1) \tilde{K}^{e^{O(2\tilde{s}-1)}} K^{e^{O(s)+O(\tilde{s})}} m} \cdot N_0, \end{aligned}$$

and symmetric sets $A_1^{(i)}, \ldots, A_{r_i}^{(i)} \subset \tilde{A}_i^{e^{O(\tilde{s}-1)}} N_i \subset \tilde{A}^{e^{O(\tilde{s})}} N_i$ such that each $A_j^{(i)} N_i$ is a $\tilde{K}^{e^{O(\tilde{s})}}$-approximate group, such that

$$[A_j^{(i)}, A_j^{(i)}] \subset N_i \tag{6.6.2}$$

for all j, and such that

$$\tilde{A}_i \subset A_1^{(i)} \cdots A_{r_i}^{(i)} N_i. \tag{6.6.3}$$

Defining $N = N_1 \cdots N_{r_0}$, we then have

$$\begin{aligned} N &\subset A^{r_0(\tilde{s}-1)\tilde{K}^{e^{O(2\tilde{s}-1)}} K^{e^{O(s)+O(\tilde{s})}m}} \cdot N_0 \\ &\subset A^{\tilde{K}^{O(1)}(\tilde{s}-1)\tilde{K}^{e^{O(2\tilde{s}-1)}} K^{e^{O(s)+O(\tilde{s})}m}} \cdot N_0 \\ &\subset A^{(\tilde{s}-1)\tilde{K}^{e^{O(2\tilde{s})}} K^{e^{O(s)+O(\tilde{s})}m}} \cdot A^{K^{e^{O(s)}m}} \\ &\subset A^{\tilde{s}\tilde{K}^{e^{O(2\tilde{s})}} K^{e^{O(s)+O(\tilde{s})}m}}. \end{aligned}$$

Since each $A_j^{(i)} N_i$ is a $\tilde{K}^{e^{O(\tilde{s})}}$-approximate group, so is each $A_j^{(i)} N$. Moreover, (6.6.1) and (6.6.3) imply that

$$\tilde{A} \subset (A_1^{(1)} \cdots A_{r_1}^{(1)}) \cdots (A_1^{(r_0)} \cdots A_{r_{r_0}}^{(r_0)}) N.$$

Finally, we have

$$\begin{aligned} r_1 + \cdots + r_{r_0} &\leq r_0 \tilde{K}^{(\tilde{s}-1)e^{O(\tilde{s})}} \\ &\leq \tilde{K}^{e^{O(\tilde{s})}} \tilde{K}^{(\tilde{s}-1)e^{O(\tilde{s})}} \\ &\leq \tilde{K}^{\tilde{s}e^{O(\tilde{s})}}, \end{aligned}$$

and $[A_j^{(i)} N, A_j^{(i)} N] \subset N$ for every i, j by (6.6.2). Taking the sets $A_j^{(i)} N$ for the sets A_k then completes the proof. □

Proof of Theorem 6.1.1 Let $N \subset A^{K^{e^{O(s)}}}$ be the normal subgroup and $A_1, \ldots, A_r \subset A^{e^{O(s)}}$ the sets given by applying Proposition 6.6.3 with $\tilde{A} = A$, noting that

$$r \leq K^{e^{O(s)}}. \tag{6.6.4}$$

Writing $\pi : G \to G/N$ for the quotient homomorphism, the proposition says that the sets $\pi(A_1), \ldots, \pi(A_r)$ generate abelian groups and are $K^{e^{O(s)}}$-approximate groups. Applying Theorem 4.1.2 modulo N therefore implies that there are subgroups $H_i \subset A_i^4 N \subset A^{K^{e^{O(s)}}}$ containing N, and ordered progressions

$$P_i \subset A_i^{K^{e^{O(s)}}} \subset A^{K^{e^{O(s)}}}, \tag{6.6.5}$$

each of rank at most $K^{e^{O(s)}}$, such that

$$A \subset H_1P_1 \cdots H_rP_r.$$

Corollary 6.5.2 then implies that for each $i = 1, \ldots, r$ there is a normal subgroup N_i of G that contains H_i and satisfies $N_i \subset A^{K^{e^{O(s)}}}$. By (6.6.4), the normal subgroup H defined by $H = N_1 \cdots N_r$ also satisfies $H \subset A^{K^{e^{O(s)}}}$, whilst (6.6.4) and (6.6.5) imply that the ordered progression P defined by $P = P_1 \cdots P_r$ satisfies $P \subset A^{K^{e^{O(s)}}}$. It follows that

$$A \subset HP \subset A^{K^{e^{O(s)}}}.$$

Moreover, (6.6.4) combined with the bound on the ranks of the P_i implies that P has rank at most $K^{e^{O(s)}}$. Writing $P = P_{\mathrm{ord}}(x; L)$, it then follows from Proposition 5.6.4 that

$$A \subset HP_{\mathrm{nil}}(x; L) \subset \overline{P}(x; L) \subset A^{K^{e^{O(s)}}},$$

as required. □

Exercises

6.1 Sanders [59] has shown that in Theorem 4.1.3 the bound on the rank of P can be improved to $O(\log^{O(1)} 2K)$, and the bound on the cardinality of $H + P$ can be improved to

$$|H + P| \geq \exp(-O(\log^{O(1)} 2K))|A|.$$

Assuming these bounds, prove the following for a finite K-approximate subgroup of an s-step nilpotent group.

(a) There exist $k \leq O(\log^{O(1)} 2K)$, a subset $X \subset A$ of size at most $\exp(O(\log^{O(1)} 2K))$, and $K^{O(1)}$-approximate groups $B_1 \ldots, B_k \subset A^{O(1)}$ such that each B_i generates a subgroup of step less than s and such that $A \subset XB_1 \cdots B_k$.

(b) There exist $\ell, m \leq O_s(\log^{O_s(1)} 2K)$, sets

$$X_1, \ldots, X_\ell \subset A^{O_s(1)}$$

of size at most $\exp(O_s(\log^{O(1)} 2K))$, and $K^{O_s(1)}$-approximate groups $C_1, \ldots, C_m \subset A^{O_s(1)}$ such that each C_i generates an abelian subgroup and such that A is contained in the product, in some order, of the X_i and C_j.

(c) There exist $n \le O_s(\log^{O_s(1)} 2K)$, sets

$$Y_1, \ldots, Y_n \subset A^{O_s(1)}$$

of size at most $\exp(O_s(\log^{O(1)} 2K))$, ordered progressions

$$P_1, \ldots, P_m \subset A^{O_s(1)}$$

of rank at most $O_s(\log^{O(1)} 2K)$, and a subgroup $H < G$ normalised by A satisfying $H \subset A^{K^{O_s(1)}}$ such that A is contained in the product, in some order, of H and the Y_i and P_j.

(d) There exist elements $y_1, \ldots, y_n \in A^{O_s(1)}$ such that there is a product, in some order, of the y_i and P_i that has size at least $\exp(-O_s(\log^{O_s(1)} 2K))|A|$.

(e) There exists an ordered progression

$$P_{\text{ord}}(x; L) \subset A^{O_s(\log^{O_s(1)} 2K)}$$

of rank at most $O_s(\log^{O_s(1)} 2K)$ such that

$$P_{\text{ord}}(x; L) \subset P_{\text{nil}}(x; L) \subset \overline{P}(x; L) \subset A^{O_s(\log^{O_s(1)} 2K)}$$

and

$$|HP_{\text{ord}}(x; L)| \ge \exp\left(-O_s(\log^{O_s(1)} 2K)\right)|AH|.$$

6.2 The aim of this question is to investigate the extent to which the results of Section 6.3 generalise to arbitrary finite nilpotent groups.

(a) Let G be a finite p-group (you may assume the well-known fact that such groups are nilpotent). Show that all minimal generating sets for G have the same size.

(b) Show that the assumption in part (a) that G is a p-group is necessary by giving an example of a finite abelian group that admits generating sets of different sizes.

(c) Let G be a finite abelian group, and let H be a subgroup of G. Show that the rank of H is at most the rank of G.

(d) Let G be a finite nilpotent group of rank r and step s. Show that if H is a subgroup of G then the rank of H is at most $f(r, s)$. Part (b) shows that we may take $f(r, 1) = r$; are there any other values of s for which we may take $f(r, s) = r$?

7
Arbitrary Approximate Groups

7.1 The Breuillard–Green–Tao Theorem

Theorem 6.1.1 showed how to classify the approximate subgroups of a nilpotent group. The following remarkable theorem essentially reduces arbitrary approximate groups to that setting.

Theorem 7.1.1 (Breuillard–Green–Tao [18]) *Let G be an arbitrary group and let $A \subset G$ be a finite K-approximate group. Then there are subgroups $H \lhd C < G$ satisfying $H \subset A^4$ such that C/H is nilpotent of step at most $O_K(1)$, and such that A is contained in the union of at most $O_K(1)$ left cosets of C.*

To make the reduction to nilpotent approximate groups more explicit, note that Lemma 2.6.2 then implies that A is covered by at most $O_K(1)$ left translates of $A^2 \cap C$, which is a K^3-approximate group by Proposition 2.6.5, and nilpotent modulo H by definition.

Since an arbitrary non-nilpotent group H is a 1-approximate group, the presence of the finite subgroup H in the conclusion of Theorem 7.1.1 is unavoidable. More generally, we make the following definition.

Definition 7.1.2 (coset nilprogression) Let P be a finite subset of a group and suppose there exists a finite subgroup $H \subset P$ that is normalised by P and such that P is a nilprogression of rank r and step s modulo H. Then P is said to be a *coset nilprogression* of rank r and step s.

It is immediate from Proposition 5.6.3 that coset nilprogressions have small tripling, at least if the parameters L_i in the definition of the nilprogression are large enough. We shall shortly see that combining Theorems 7.1.1 and 6.1.1 shows that coset nilprogressions are essentially the only examples of sets of small tripling, as follows.

Theorem 7.1.3 (Breuillard–Green–Tao [18]) *Let G be an arbitrary group and let $A \subset G$ be a finite K-approximate group. Then there exists a coset nilprogression P of rank and step at most $O_K(1)$ such that A is covered by at most $O_K(1)$ left translates of P.*

The deduction of Theorem 7.1.3 from Theorem 7.1.1 is captured by the following proposition.

Proposition 7.1.4 (deduction of Theorem 7.1.3 from Theorem 7.1.1) *Let G be an arbitrary group and let $A \subset G$ be a finite K-approximate group. Suppose that $H \lhd C < G$ and that C/H is nilpotent of step s, and that A is contained in the union of at most k left cosets of C. Then there is a set $P \subset A^{K^{e^{O(s)}}}$ that is a coset nilprogression of rank $K^{e^{O(s)}}$ modulo H such that A is covered by at most k left translates of HP.*

Note that if the group H in Proposition 7.1.4 is finite then HP is a coset nilprogression of rank $K^{e^{O(s)}}$ and step s.

Proof Lemma 2.6.2 implies that A is covered by at most k left translates of $A^2 \cap C$, which is a K^3-approximate group by Proposition 2.6.5. The image of $A^2 \cap C$ modulo H is therefore also a K^3-approximate group, and so the proposition follows from applying Theorem 6.1.1 to this image. □

A proof of Theorem 7.1.1 is beyond the scope of this book. In particular, it uses tools from model theory that are quite different from the material of this book. We will nonetheless assume Theorem 7.1.1 in Chapter 11 in order to demonstrate some of its applications.

Breuillard, Green and Tao originally proved Theorem 7.1.3 directly, without passing through Theorem 7.1.1 or using Proposition 7.1.4; indeed, Theorem 7.1.3 pre-dates Theorem 6.1.1, which is the main content of Proposition 7.1.4.

There are nonetheless good reasons for presenting Theorem 7.1.3 as a corollary of Theorem 7.1.1 and Proposition 7.1.4. The first relates to the so-called *effectiveness* of the bounds in Theorem 7.1.3. A bound on a quantity is often said to be *effective* if it is given explicitly, or at least could be if one kept track of all of the bounds in an argument. A bound that is not effective is said to be *ineffective*. All except one of the constants in Theorem 7.1.1, and hence Theorem 7.1.3, can be made effective in terms of K. The exception is the bound on the number of cosets of C needed to cover A, for which there is no known explicit computation. Theorems 7.1.1 and 7.1.3 thus do not quite represent the final word on the subject that they might appear to at first glance. Proposition 7.1.4 means that in order to prove an effective version of Theorem 7.1.3 it would be sufficient to establish the a priori weaker Theorem 7.1.1 effectively.

The second, related, reason is that one can make Theorem 7.1.1 effective at the expense of restricting attention to certain specific classes of groups. This is notably the case for various linear groups, and for *residually nilpotent* groups, for example. We will illustrate this in Chapters 8–10, where we will prove Theorem 7.1.1 effectively, first in the case where G is residually nilpotent, then when G is a *soluble* subgroup of $GL_n(\mathbb{C})$, and finally for $G = GL_n(\mathbb{C})$ itself. By Proposition 7.1.4, each of these results also leads to an effective version of Theorem 7.1.3 for the class of groups concerned.

For the results of Chapters 8–10 it will be useful to remark that when proving Theorem 7.1.1, instead of showing that A is contained in a union of a few left cosets of C, it is sufficient to show that some small power of A has large intersection with C, as follows.

Lemma 7.1.5 *Let G be a group with a subgroup C, and let $A \subset G$ be a finite K-approximate group. Let $\alpha > 0$, and suppose that $|A^m \cap C| \geq \alpha|A|$. Then A is contained in the union of at most $\alpha^{-1}K^m$ left cosets of the subgroup C.*

Proof Writing $\pi : G \to G/C$ for the quotient homomorphism, we have

$$\begin{aligned} |\pi(A)| &\leq \frac{|A^{m+1}|}{|A^m \cap C|} && \text{(by Lemma 2.6.3)} \\ &\leq \alpha^{-1}\frac{|A^{m+1}|}{|A|} \\ &\leq \alpha^{-1}K^m && \text{(since } A \text{ is a } K\text{-approximate group).} \end{aligned}$$

This means precisely that A is contained in the union of at most $\alpha^{-1}K^m$ left cosets of C, as required. □

8
Residually Nilpotent Approximate Groups

8.1 Introduction

In this chapter we prove Theorem 7.1.1 for *residually nilpotent* groups. Given a property $\mathcal{P}$ of groups, a group G is said to be *residually* $\mathcal{P}$ if for every non-identity element $g \in G$ there exists a group Γ with property $\mathcal{P}$ and a homomorphism $\pi : G \to \Gamma$ such that $\pi(g) \neq 1$. In particular, G is *residually nilpotent* if for every non-identity element $g \in G$ there exists a nilpotent group N and a homomorphism $\pi : G \to N$ such that $\pi(g) \neq 1$.

Being residually nilpotent is a strictly weaker condition than that of being nilpotent: finitely generated non-abelian free groups are residually nilpotent but not nilpotent, for example.

The main result of this chapter is the following.

Theorem 8.1.1 ([73, Theorem 1.2]) *Let G be a residually nilpotent group and suppose that A is a finite K-approximate subgroup of G. Then there exist subgroups $H \lhd C < G$ satisfying $H \subseteq A^{O_K(1)}$ such that C/H is nilpotent of step at most K^6, and such that A is contained in the union of at most $\exp(K^{O(1)})$ left cosets of C.*

In light of Proposition 7.1.4, this also gives the following.

Theorem 8.1.2 ([73, Corollary 1.4]) *Let G be a residually nilpotent group and suppose that A is a finite K-approximate subgroup of G. Then there exists a coset nilprogression $P \subset A^{O_K(1)}$ of rank at most $\exp(\exp(K^{O(1)}))$ and step at most K^6 such that A is covered by at most $\exp(K^{O(1)})$ left translates of P.*

Note in particular that every nilpotent group is also residually nilpotent, and so Theorem 8.1.2 implies a version of Theorem 6.1.1 in which the bounds are independent of the step of G.

A version of Theorem 8.1.1 was first proved for the special case of residually torsion-free nilpotent groups by Breuillard, Green and Tao [19]. The proof we give here of Theorem 8.1.1 in general was originally given by the author in [73].

We start by proving a version of Theorem 8.1.1 in the special case where G is nilpotent, in the form of the following proposition.

Proposition 8.1.3 *Let G be a nilpotent group and suppose that A is a finite K-approximate subgroup of G. Then there exist subgroups $H \lhd C < G$ such that*

(i) $H \subseteq A^{O_K(1)}$;
(ii) C/H is nilpotent of step at most K^6;
(iii) $|A^2 \cap C| \geq \exp(-K^{O(1)})|A|$;
(iv) C is generated by $A^6 \cap C$.

We prove Proposition 8.1.3 over the next three sections. Then, in Section 8.5, we deduce Theorem 8.1.1 from it.

8.2 Central Extensions of Nilpotent Approximate Groups

We start our proof of Proposition 8.1.3 in the setting of 'central extensions' of nilpotent approximate groups. A group G is a *central extension* of a group H if there is a central subgroup $Z < G$ such that $G/Z \cong H$. Here we analogously refer to G as a *central extension* of an approximate group $A \subset G$ if there is a central subgroup $Z < G$ such that $G = AZ$.

Proposition 8.2.1 *Let G be a finitely generated nilpotent group and let A be a finite K-approximate subgroup. Suppose there exists a central subgroup $Z < G$ such that $G = AZ$. Then there exist $k \leq K^8$ and normal subgroups $\{1\} = H_0 < H_1 < \cdots < H_k < [G, G]$ of G such that $H_i \subset A^8 H_{i-1}$, and such that $[G, G] \subset A^4 H_k$. In particular, $[G, G] \subseteq A^{8K^8+4}$.*

In Exercise 8.1 we invite the reader to prove a strong version of Theorem 8.1.2 for central extensions of nilpotent approximate groups.

For the remainder of this section G is a finitely generated nilpotent group. Fix a generating set $x_1, \ldots, x_r$ for G, and write $c_1, \ldots, c_q$ for the set of simple commutators in the x_i of total weight at least 2 and at most s, in non-increasing order of total weight. For each $i = 0, 1, \dots, q$, define the subgroup

$$\Gamma_i = \langle c_1, \dots, c_i \rangle,$$

noting from Proposition 5.2.6 and Corollary 5.2.9 that $\Gamma_q = [G, G]$, and from Propositions 5.2.4 and 5.2.6 and Corollary 5.2.9 that $\{1\} = \Gamma_0 < \Gamma_1 < \cdots < \Gamma_q = [G, G]$ is a central series for $[G, G]$. Note also therefore that

$$\Gamma_i = \langle c_i \rangle \Gamma_{i-1} \tag{8.2.1}$$

for $i = 1, \dots, q$. We keep these commutators c_i and subgroups Γ_i fixed for the remainder of this section.

The rough idea of the proof of Proposition 8.2.1, which also plays an important role in our subsequent proof of Theorem 8.1.1, is to view each Γ_i/Γ_{i-1} as a different 'direction' in G, and then to show that A cannot grow in too many of these different directions. This is the idea behind Freiman's so-called dimension lemma, which states that a finite set of doubling K inside $\mathbb{R}^d$ is contained in an affine subspace of dimension at most $K-1$ [68, Theorem 5.20]. The following lemma, which goes back to Gleason [29] and was first applied to approximate groups by Breuillard, Green and Tao [19], can be thought of as a generalisation of this idea. Roughly, it shows that if a set grows even a little bit in several different 'directions' in a group then it must grow a lot overall.

Lemma 8.2.2 *Let A be a finite symmetric subset of a group and let $m \in \mathbb{N}$. Let $\{1\} = H_0 < H_1 < \cdots < H_k$ be a nested sequence of groups such that $A^m \cap H_i \not\subseteq A^2 H_{i-1}$. Then $|A^{m+1}| \geq k|A|$.*

Proof This is essentially [19, Lemma 3.1]. For each $i = 1, \dots, k$ pick $h_i \in (A^m \cap H_i) \setminus A^2 H_{i-1}$. It is sufficient to show that the sets Ah_i are all disjoint. To see this, suppose that $Ah_i \cap Ah_j \neq \varnothing$ for some $j < i$. This would imply that $h_i \in A^2 h_j \subseteq A^2 H_j \subseteq A^2 H_{i-1}$, contradicting the choice of h_i. □

We spend the rest of this section showing how to use Lemma 8.2.2 to prove Proposition 8.2.1. To that end, for the remainder of the section we fix a finite K-approximate group $A \subset G$ and a central subgroup $Z < G$ such that $G = AZ$ as in Proposition 8.2.1.

Lemma 8.2.3 *Let $x, y \in G$. Then $[x, y] \in A^4$.*

Proof Since $G = AZ$ there exist $a, b \in A$ and $w, z \in Z$ such that $x = aw$ and $y = bz$. Since Z is central, we therefore have $[x, y] = [aw, bz] = [a, b] \in A^4$, as required. □

Lemma 8.2.4 *For each $i = 1, \ldots, r$ we have $\Gamma_i \subseteq A^4\Gamma_{i-1}$.*

Proof Write k for the weight of c_i. By definition we have $c_i = [d, x_j]$ for some j and some simple commutator d in the x_ℓ of weight $k-1$. It follows from Proposition 5.2.6 and Corollary 5.2.9 that $G_{k+1} \subset \Gamma_{i-1}$, and hence from the last part of Lemma 5.5.3 that $c_i^\ell = [d, x_j]^\ell \in [d, x_j^\ell]\Gamma_{i-1}$ for every $\ell \in \mathbb{Z}$. The desired result therefore follows from (8.2.1) and Lemma 8.2.3. □

Lemma 8.2.5 *Let $j \in \{0, \ldots, r-1\}$. Then there exists $j' > j$ such that $\Gamma_{j'} \subseteq A^4\Gamma_j$, and such that either $j' = r$ or $\Gamma_{j'+1} = A^8\Gamma_j \cap \Gamma_{j'+1} \not\subseteq A^4\Gamma_j$.*

Proof Let $j' \leq r$ be maximal such that $(A^4 \cap \Gamma_{j'})\Gamma_j$ is a group, noting that $j' > j$ by Lemma 8.2.4. Lemma 8.2.4 implies that $(A^4 \cap \Gamma_{j'})$ generates $\Gamma_{j'}$, so in fact we have $(A^4 \cap \Gamma_{j'})\Gamma_j = \Gamma_{j'}$; in particular,

$$\Gamma_{j'} \subseteq A^4\Gamma_j, \tag{8.2.2}$$

as required. If $j' \neq r$ then $(A^4 \cap \Gamma_{j'+1})\Gamma_j$ is not a group by definition of j', and in particular we have $(A^8 \cap \Gamma_{j'+1})\Gamma_j \not\subseteq (A^4 \cap \Gamma_{j'+1})\Gamma_j$, and hence $A^8\Gamma_j \cap \Gamma_{j'+1} \not\subseteq A^4\Gamma_j$. However, $\Gamma_{j'+1} = A^8\Gamma_j \cap \Gamma_{j'+1}$ by (8.2.2) and Lemma 8.2.4. □

Proof of Proposition 8.2.1 Repeated application of Lemma 8.2.5 implies that there exist $k \in \mathbb{Z}$ and $0 = j(0) < j(1) < \cdots < j(k)$ such that $\Gamma_{j(i)} = A^8\Gamma_{j(i-1)} \cap \Gamma_{j(i)} \not\subseteq A^4\Gamma_{j(i-1)}$ for each i, and such that $[G, G] \subset A^4\Gamma_{j(k)}$. Lemma 8.2.2 implies that $k \leq K^8$, and so we may take $H_i = \Gamma_{j(i)}$ in Proposition 8.2.1. □

8.3 Bounded Normal Series for Nilpotent Approximate Groups

Proposition 8.2.1 roughly states that if G is a central extension of a finite nilpotent approximate group A then there exist normal subgroups $\{1\} = H_0 < H_1 < \cdots < H_k$ of $[G, G]$ with

$$H_i \subset A^8 H_{i-1} \tag{8.3.1}$$

and $k \leq K^8$ such that $[G, G] \subset A^4 H_k$. The property (8.3.1) can be thought of as saying that the extension of $[G, G]/H_i$ by H_i/H_{i-1} is 'bounded' in terms of A, and so Proposition 8.2.1 can be thought of as writing $[G, G]$ as a bounded series of 'bounded group extensions'.

Unsurprisingly, such a series does not exist in an arbitrary nilpotent group. Nonetheless, we have the following analogue that does exist in general.

Proposition 8.3.1 *Let G be a nilpotent group and let $A \subset G$ be a finite K-approximate group. Then there exists a normal series*

$$\{1\} = H_0 \lhd D_1 \lhd H_1 \lhd D_2 \lhd \cdots \lhd D_k \lhd H_k \lhd D_{k+1} \tag{8.3.2}$$

inside G, with each $H_i = \langle \gamma_i \rangle D_i$ for some element $\gamma_i \in G$, such that

(a) $D_i \subset A^2 H_{i-1}$ for each i;
(b) $\gamma_i \in A^6 \setminus A^2 H_{i-1}$ for each $i \geq 1$;
(c) $D_i \lhd D_{k+1}$ for each i;
(d) H_i is central modulo D_i in D_{k+1} for each i;
(e) $|A^2 \cap D_{k+1}| \geq K^{-30K^6}|A|$;
(f) $k \leq K^6$.

Rather than a bounded series of bounded group extensions as in Proposition 8.2.1, (8.3.2) can be thought of as writing D_{k+1} as a bounded series of some bounded extensions and some central extensions. Specifically, each D_{k+1}/H_i is a bounded extension of D_{k+1}/D_{i+1} by D_{i+1}/H_i (bounded in the sense that $D_{i+1} \subset A^2 H_i$ by (a)), whilst each D_{k+1}/D_i is a central extension of D_{k+1}/H_i by H_i/D_i.

A nilpotent group of bounded step is by definition a group that can be written as a bounded series of central extensions, and so Proposition 8.1.3 can be thought of as writing a nilpotent approximate group as a bounded series of central extensions, followed by a single bounded extension. Proposition 8.3.1 thus points in the direction of Proposition 8.1.3, the issue being that the series (8.3.2) has bounded extensions throughout, rather than just at the left-hand end.

In the next section we will explain how to disentangle the series (8.3.2) and prove Proposition 8.1.3. For now we prove Proposition 8.3.1, starting with some preliminary lemmas, as follows.

Lemma 8.3.2 *Let G be a group with symmetric generating set B containing the identity, and let Γ be a normal subgroup of G. Suppose that $B^3 \cap \Gamma \subset B$. Then $B \cap \Gamma$ is a normal subgroup of G.*

Proof We have $(B\cap\Gamma)^2 \subset B^2\cap\Gamma \subset B\cap\Gamma$, so $B\cap\Gamma$ is a group. Moreover, given $b \in B$ and $g \in B\cap\Gamma$, we have $b^{-1}gb \in B^3\cap\Gamma \subset B\cap\Gamma$, so B normalises $B\cap\Gamma$. Since B generates G, the subgroup $B\cap\Gamma$ is therefore normal in G. □

The following lemma was first applied in this context by Breuillard, Green and Tao [19]; it is also somewhat reminiscent of [39, Proposition 4.1].

Lemma 8.3.3 *Let $m \in \mathbb{N}$. Suppose that G is a group with subgroups $H_1, H_2 < G$ satisfying $[G, H_1] \subset H_2$. Suppose further that $A \subset G$ is a finite K-approximate group satisfying $A^2 \cap H_2 = \{1\}$. Let $\omega \in A^m \cap H_1$. Then*

$$|A^2 \cap C_G(\omega)| \geq \frac{|A|}{K^{2m+1}}.$$

Proof This is essentially found in the proof of [19, Proposition 4.1]. Proposition 2.6.5 implies that $A^{2m+2}\cap H_2$ is covered by K^{2m+1} translates of $A^2 \cap H_2$, and hence that $|A^{2m+2} \cap H_2| \leq K^{2m+1}$. Since every $a \in A$ satisfies $[\omega, a] \in A^{2m+2} \cap H_2$, this implies in particular that as a ranges through A the number of values taken by $[\omega, a]$ is at most K^{2m+1}. There therefore exists a such that $[\omega, x] = [\omega, a]$ for at least $|A|/K^{2m+1}$ elements $x \in A$. For each such x, by definition we have $\omega^{-1}x^{-1}\omega x = \omega^{-1}a^{-1}\omega a$, hence $x^{-1}\omega x = a^{-1}\omega a$, hence $ax^{-1}\omega xa^{-1} = \omega$, and hence $xa^{-1} \in A^2 \cap C_G(\omega)$. □

Finally, it is convenient to separate out the following straightforward lemma, since we use it on two separate occasions in the proof of Proposition 8.3.1.

Lemma 8.3.4 *Let G be a group with subgroups $H < U < G$, and let $A \subset G$. Then $AH \cap U = (A \cap U)H \cap U$.*

Proof The fact that $AH \cap U \supset (A \cap U)H \cap U$ is trivial. To see that $AH \cap U \subset (A \cap U)H \cap U$, suppose that $u \in AH \cap U$, so that there exist $a \in A$ and $h \in H$ such that $u = ah$. We then have $a = uh^{-1} \in U$, and hence $u \in (A \cap U)H$. □

Proof of Proposition 8.3.1 To facilitate a recursive construction of the series (8.3.2) we will also construct a series

$$\langle A\rangle = C_0 > C_1 > C_2 > \cdots > C_k = D_{k+1} \tag{8.3.3}$$

such that

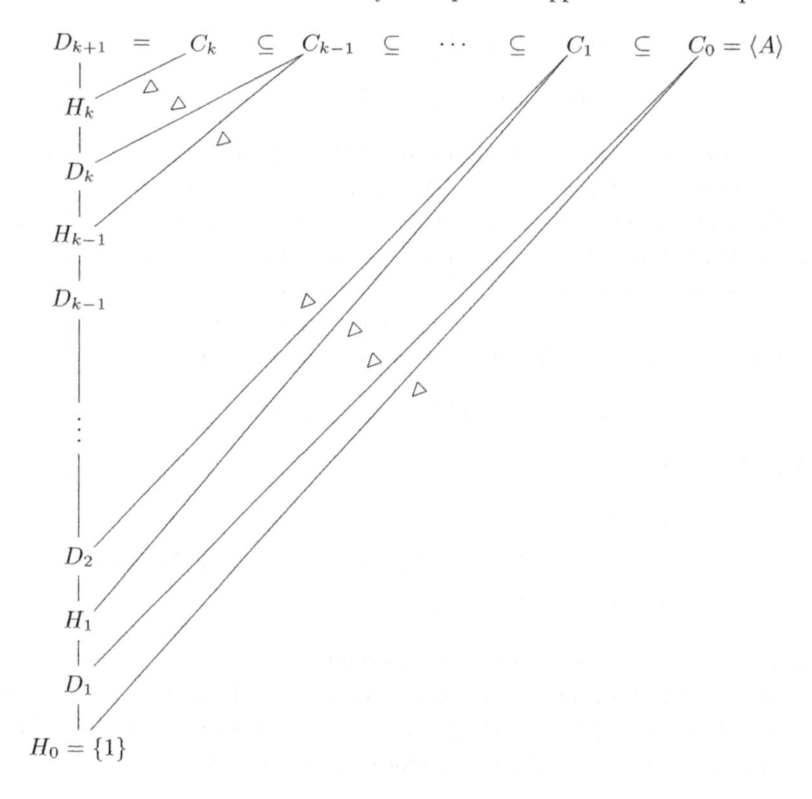

Figure 8.1 The subgroups appearing in the proof of Proposition 8.3.1

(1) $D_i \lhd C_{i-1}$;
(2) γ_i is in the centre of C_i modulo D_i;
(3) $C_i = \langle A^2 \cap C_i \rangle \langle \gamma_i \rangle D_i$;
(4) $|A^2 \cap C_i| \geq K^{-30i}|A|$.

Setting $H_i = \langle \gamma_i \rangle D_i$ for each $i \geq 1$, the subgroups H_i and D_i will then automatically be normal in D_{k+1}, so (8.3.2) will be a normal series, as claimed. Moreover, conclusion (f) will follow from (b) and Lemma 8.2.2, whilst (e) will follow from (8.3.3), (f) and (4). On the other hand, (c) and (d) will follow from (8.3.3), (b), (1) and (2). It is sufficient, therefore, to construct elements γ_i and subgroups C_i, D_i and $H_i = \langle \gamma_i \rangle D_i$ satisfying (a), (b) and (1)–(4). The relationships between the subgroups C_i, D_i and H_i are illustrated in Figure 8.1.

We construct the elements γ_i and subgroups C_i, D_i and $H_i = \langle \gamma_i \rangle D_i$ recursively, starting with $H_0 = \{e\}$ and $C_0 = \langle A \rangle$. Suppose, then, that we have already constructed subgroups

$$H_0,\dots,H_j,D_1,\dots,D_j,C_0,\dots,C_j$$

and elements $\gamma_1,\dots,\gamma_j$ satisfying the required properties. If $C_j\subset A^2H_j$ then set $D_{j+1}=C_j$ and $k=j$, and the proposition is proved.

We may suppose, therefore, that $C_j\not\subset A^2H_j$. Write $C_j=Y_1\rhd\cdots\rhd Y_m=\{1\}$ for the lower central series of C_j, and set $Z_i=Y_iH_j$ for each i. Let ℓ be maximal such that $((A^2\cap C_j)^3\cap Z_\ell)\backslash(A^2\cap C_j)H_j\ne\varnothing$, noting that such an ℓ exists since $A^2\cap C_j$ generates C_j, which by assumption is not contained in A^2H_j. We may then fix an arbitrary

$$\gamma_{j+1}\in\big((A^2\cap C_j)^3\cap Z_\ell\big)\setminus(A^2\cap C_j)H_j.$$

Note in particular that

$$\begin{aligned}\gamma_{j+1}&\in\big((A^2\cap C_j)^3\cap Z_\ell\big)\setminus\big((A^2\cap C_j)H_j\cap C_j\big)\\&\subset\big((A^2\cap C_j)^3\cap Z_\ell\big)\setminus\big(A^2H_j\cap C_j\big)&&\text{(by Lemma 8.3.4)}\\&\subset\big((A^2\cap C_j)^3\cap Z_\ell\big)\setminus A^2H_j,\end{aligned}$$

so that γ_{j+1} satisfies property (b), as required.

By the definition of ℓ and Lemma 8.3.2 applied modulo H_j, the set $\big((A^2\cap C_j)\cap Z_{\ell+1}\big)H_j=(A^2\cap Z_{\ell+1})H_j$ is a normal subgroup of C_j. The set $D_{j+1}=(A^2\cap Z_{\ell+1})H_j$ is therefore a group satisfying properties (a) and (1).

Let n be maximal such that $(A^2\cap C_j)^2\cap Z_n\not\subset D_{j+1}$. By definition of D_{j+1}, this means that there exists

$$z\in\big((A^2\cap C_j)^3\cap Z_n\big)\setminus(A^2\cap Z_{\ell+1})H_j.$$

If $n>\ell$ this would mean that

$$\begin{aligned}z&\in\big((A^2\cap C_j)^3\cap Z_{\ell+1}\big)\setminus(A^2\cap Z_{\ell+1})H_j\\&\subset\big((A^2\cap C_j)^3\cap Z_{\ell+1}\big)\setminus\big(((A^2\cap C_j)\cap Z_{\ell+1})H_j\cap Z_{\ell+1}\big)\\&\subset\big((A^2\cap C_j)^3\cap Z_{\ell+1}\big)\setminus\big((A^2\cap C_j)H_j\cap Z_{\ell+1}\big)\end{aligned}$$

by Lemma 8.3.4, and hence

$$z\in\big((A^2\cap C_j)^3\cap Z_{\ell+1}\big)\setminus(A^2\cap C_j)H_j,$$

contradicting the definition of ℓ. It follows that $n\le\ell$, and hence that

$$\gamma_{j+1}\in(A^2\cap C_j)^3\cap Z_n. \tag{8.3.4}$$

Write $\pi_j:C_j\to C_j/D_{j+1}$ for the quotient homomorphism, and note that $[C_j/D_{j+1}, \pi_j(Z_n)] \subset \pi(Z_{n+1})$ by Proposition 5.2.4 and the definition of the groups Z_i. Since $A^2 \cap C_j$ is a K^3-approximate group by Proposition 2.6.5, (8.3.4) therefore implies that applying Lemma 8.3.3 modulo D_{j+1} gives

$$\left|\pi_j((A^2 \cap C_j)^2) \cap C_{C_j/D_{j+1}}(\gamma_{j+1})\right| \geq K^{-19} \left|\pi_j(A^2 \cap C_j)\right|. \qquad (8.3.5)$$

Writing $C'_{j+1} = \pi_j^{-1}(C_{C_j/D_{j+1}}(\gamma_{j+1}))$, we claim that

$$|A^2 \cap C'_{j+1}| \geq K^{-30}|A^2 \cap C_j|. \qquad (8.3.6)$$

Lemma 2.6.2 implies that

$$|A^2 \cap C_j| \leq |\pi_j(A^2 \cap C_j)||(A^2 \cap C_j)^2 \cap D_{j+1}|. \qquad (8.3.7)$$

We also have

$$\left|\pi_j((A^2 \cap C_j)^2) \cap C_{C_j/D_{j+1}}(\gamma_{j+1})\right| \subset \pi_j(A^4 \cap C'_{j+1}),$$

which combines with (8.3.5) to imply that

$$|\pi_j(A^2 \cap C_j)| \leq K^{19}|\pi_j(A^4 \cap C'_{j+1})|. \qquad (8.3.8)$$

The fact that $(A^2 \cap C_j)^2 \cap D_{j+1} \subset A^4 \cap D_{j+1} \subset A^4 \cap C'_{j+1} \cap D_{j+1}$ implies that

$$|(A^2 \cap C_j)^2 \cap D_{j+1}| \leq |(A^4 \cap C'_{j+1})^2 \cap D_{j+1}|. \qquad (8.3.9)$$

By Lemma 2.6.3 we also have

$$\begin{aligned} |\pi_j(A^4 \cap C'_{j+1})||(A^4 \cap C'_{j+1})^2 \cap D_{j+1}| &\leq |(A^4 \cap C'_{j+1})^3| \\ &\leq |A^{12} \cap C'_{j+1}|, \end{aligned}$$

and hence

$$|\pi_j(A^4 \cap C'_{j+1})||(A^4 \cap C'_{j+1})^2 \cap D_{j+1}| \leq K^{11}|(A^2 \cap C'_{j+1})| \qquad (8.3.10)$$

by Proposition 2.6.5. Combining (8.3.7), (8.3.8), (8.3.9) and (8.3.10) proves (8.3.6), as claimed.

Setting $H_{j+1} = \langle\gamma_{j+1}\rangle D_{j+1}$ and $C_{j+1} = \langle A^2 \cap C'_{j+1}\rangle H_{j+1}$, the subgroup C_{j+1} satisfies (4) by (8.3.6) and induction, and satisfies (3) by definition. Since $C_{j+1} \subset C'_{j+1}$, the element γ_{j+1} also satisfies (2) by definition of C'_{j+1}. This completes the proof. □

8.4 From Normal to Central Subgroups

The aim of this section is to convert the bounded normal series (8.3.2) given by Proposition 8.3.1 into a bounded extension of a bounded central series, and thus prove Proposition 8.1.3.

We use the following special case of a lemma of Guralnick [37], in which, given a subgroup D and an element x of a group G, we write $[x, D]$ for the set $\{[x, d] : d \in D\}$.

Lemma 8.4.1 ([37, Lemma 3.1]) *Let G be a group, and let D be an abelian normal subgroup of G such that $G = \langle x_1, \ldots, x_n, D\rangle$. Then $[G, D] = \prod_{i=1}^{n}[x_i, D]$.*

Proof Since D is normal we have $[G, D] \subset D$, and in particular commutes with D. Lemma 5.5.3 therefore implies that $[x_i, d][x_i, d'] = [x_i, dd']$ for each i and every $d, d' \in D$, and hence that each $[x_i, D]$ is a subgroup of D. Since D is abelian, this implies in turn that $\prod_{i=1}^{n}[x_i, D]$ is a subgroup. Furthermore, for every i, j and $d \in D$ we have $x_j^{-1}[x_i, d]x_j = [x_j, [x_i, d]][x_i, d]$, and so $\prod_{i=1}^{n}[x_i, D]$ is in fact a normal subgroup.

We may therefore consider the quotient $G/\prod_{i=1}^{n}[x_i, D]$. Since D is central in this quotient, the subgroup $[G, D]$ is trivial in this quotient, so it must be that $[G, D] \subset \prod_{i=1}^{n}[x_i, D]$. The converse inclusion is trivial, and so the lemma is proved. □

Proof of Proposition 8.1.3 Let $k, D_1, \ldots, D_{k+1}, \gamma_1 \ldots, \gamma_k$ be as given by Proposition 8.3.1, and set $C = D_{k+1}$. Proposition 8.3.1 (e) implies that $|A^2 \cap C| \geq K^{-30K^6}|A| \geq \exp(-K^{O(1)})|A|$, giving (iii). Meanwhile, (iv) follows from Proposition 8.3.1 (a) and (b).

We will exhibit groups $\overline{D}_1 \lhd \overline{D}_2 \lhd \cdots \lhd \overline{D}_k \lhd \overline{D}_{k+1} = C$ with $D_i \subseteq \overline{D}_i \subseteq A^{O_{K,k-i}(1)}D_i$, and each $\overline{D}_i$ normal in C, such that $\overline{D}_{i+1}$ is central in C modulo $\overline{D}_i$. Setting $H = \overline{D}_1$, we will then have $H \subset A^{O_K(1)}$ by Proposition 8.3.1 (a) and (f), giving (i). Moreover, C/H will be nilpotent of step at most k, which is at most K^6 by Proposition 8.3.1 (f), giving (ii) and completing the proof of the proposition.

To facilitate the construction of these groups $\overline{D}_i$, we also construct elements

$$\begin{matrix} x_{1,1} & \cdots & x_{1,r_1} \\ \vdots & & \vdots \\ x_{k,1} & \cdots & x_{k,r_k} \end{matrix} \in A^{O_{K,k-i}(1)}$$

with each $r_i \ll_{K,k-i} 1$ such that $\overline{D}_{i+1} = \langle x_{i,1}, \ldots, x_{i,r_i}, \overline{D}_i\rangle$ for every i.

We construct the elements $x_{i,j}$ and the groups $\overline{D}_i$ recursively, starting by setting $\overline{D}_{k+1} = C$. Suppose, then, that we have already constructed groups $\overline{D}_{i+1}, \ldots, \overline{D}_{k+1}$ and elements

$$\begin{matrix} x_{i+1,1} & \cdots & x_{i+1,r_{i+1}} \\ \vdots & & \vdots \\ x_{k,1} & \cdots & x_{k,r_k} \end{matrix}$$

satisfying the required properties. This means in particular that each $x_{j,\ell} \in A^R$ for some $R \ll_{K,k-i-1} 1$, and $\overline{D}_{i+1} \subset A^M D_{i+1}$ for some $M \ll_{K,k-i-1} 1$. Proposition 8.3.1 (a) and (b) then imply that

$$\overline{D}_{i+1} \subset A^{M+2} \langle \gamma_i \rangle D_i. \tag{8.4.1}$$

Proposition 2.6.5 implies that $A^{M+2} \cap \overline{D}_{i+1}$ is a K^{2M+3}-approximate group. Since γ_i is central in C modulo D_i by Proposition 8.3.1 (b) and (d), (8.4.1) means that we may apply Proposition 8.2.1 modulo D_i to conclude that

$$[\overline{D}_{i+1}, \overline{D}_{i+1}] \subset A^{8K^{16M+24}+4} D_i. \tag{8.4.2}$$

Now by assumption we have

$$\left\langle \begin{matrix} x_{i+1,1} & \cdots & x_{i+1,r_{i+1}} \\ \vdots & & \vdots \\ x_{k,1} & \cdots & x_{k,r_k} \end{matrix}, \overline{D}_{i+1} \right\rangle,$$

and modulo $[\overline{D}_{i+1}, \overline{D}_{i+1}] D_i$ the group $\overline{D}_{i+1}$ is abelian, so Lemma 8.4.1 implies that

$$[C, \overline{D}_{i+1}] \subset \left(\prod_{j=i+1}^{k} \prod_{\ell=1}^{r_i} [x_{j,\ell}, \overline{D}_{i+1}] \right) [\overline{D}_{i+1}, \overline{D}_{i+1}] D_i. \tag{8.4.3}$$

However, since γ_i is central modulo D_i, for every $x_{j,\ell}$ and $d \in \overline{D}_{i+1}$ the image of $[x_{j,\ell}, d]$ modulo D_i depends only on the image of d modulo $\langle \gamma_i \rangle D_i$. Combined with (8.4.1) and the fact that $x_{j,\ell} \in A^R$, this means that $[x_{j,\ell}, d] \subset A^{2M+2R+4} D_i$. Combined with (8.4.2) and (8.4.3), this in turn means that

$$[C, \overline{D}_{i+1}] \subset A^{O_{K,k-i}(1)} D_i. \tag{8.4.4}$$

Now $\gamma_i \in A^6$ by Proposition 8.3.1 (b), so (8.4.1) implies that $\overline{D}_{i+1}$ is generated by D_i and $A^{M+2} \cap \overline{D}_{i+1}$. Since $\overline{D}_{i+1}$ is normal in C, so is $[C, \overline{D}_{i+1}]$, and $\overline{D}_{i+1}$ is abelian modulo $[C, \overline{D}_{i+1}]$. Since $A^{M+2} \cap \overline{D}_{i+1}$ is a K^{2M+3}-approximate group by Proposition 2.6.5, we may therefore apply Theorem 4.1.2 to $A^{M+2} \cap \overline{D}_{i+1}$ modulo $[C, \overline{D}_{i+1}] D_i$ to conclude that there exist $r_i \le K^{O(M)} \ll_{K,k-i} 1$, elements $x_{i,1},\dots,x_{i,r_i} \in A^{4M+8}$, and a subgroup $\overline{D}_i < \overline{D}_{i+1}$ satisfying

$$[C,\overline{D}_{i+1}]D_i \subset \overline{D}_i \subset A^{4M+8}[C,\overline{D}_{i+1}]D_i \tag{8.4.5}$$

such that

$$\overline{D}_{i+1} = \langle x_{i,1},\dots,x_{i,r_i},\overline{D}_i\rangle.$$

Since $\overline{D}_i < \overline{D}_{i+1}$, it is central modulo $[C,\overline{D}_{i+1}]$, and hence normal in C. Moreover, since $[C,\overline{D}_{i+1}] \subset \overline{D}_i$ we have $\overline{D}_{i+1}$ central in C modulo $\overline{D}_i$. Finally, (8.4.5) and (8.4.4) imply that $\overline{D}_i \subset A^{O_{K,k-i}(1)}D_i$, as required, and so the proposition is proved. □

8.5 Residually Nilpotent Groups

In this section we complete the proof of Theorem 8.1.1. We use an argument first deployed by Breuillard, Green and Tao in the torsion-free case to reduce Theorem 8.1.1 to the case in which G is nilpotent, and hence to Proposition 8.1.3.

We first note that being residually nilpotent is in fact equivalent to an apparently slightly stronger condition, as follows.

Lemma 8.5.1 *Let G be a residually nilpotent group and let $A \subseteq G$ be a finite set such that $1 \notin A$. Then there exists a nilpotent group N and a homomorphism $\pi : G \to N$ such that $A \cap \ker\pi = \varnothing$.*

Proof By definition, for each $a \in A$ there exists a nilpotent group N_a and a homomorphism $\pi_a : G \to N_a$ such that $\pi_a(a) \ne 1$. In particular, writing s_a for the step of N_a and $G = G_1 \supseteq G_2 \supseteq \cdots$ for the lower central series of G we have $a \notin G_{s_a+1}$. Writing $s = \max_{a\in A} s_a$, we may therefore take π to be the projection homomorphism $\pi : G \to G/G_{s+1}$. □

Lemma 8.5.2 *Let G be a group, and let $A \subseteq G$ be a symmetric set containing the identity. Let N be another group, and let $\pi : G \to N$ be a homomorphism. Let $H \subset \pi(A)$ be a subgroup of N. Then we have the following.*

(i) *If $A^2 \cap \ker\pi = \{1\}$ then π is injective on A.*
(ii) *If $A^3 \cap \ker\pi = \{1\}$ then there exists a subgroup $H' \subseteq A$ isomorphic to H via π.*

(iii) If $A^4 \cap \ker\pi = \{1\}$ then H' is normal in $\langle A\rangle$ if and only if H is normal in $\langle\pi(A)\rangle$.

Proof Item (i) follows from the fact that if $\pi(a) = \pi(a')$ then $a^{-1}a' \in \ker\pi$, and in turn implies that for each $h \in H$ there is a unique $\phi(h) \in A$ such that $\pi(\phi(h)) = h$. Given $h, h' \in H$ we have $\phi(h)\phi(h')\phi(hh')^{-1} \in A^3 \cap \ker\pi$. If $A^3 \cap \ker\pi = \{1\}$, it therefore follows that $\phi(h)\phi(h') = \phi(hh')$, and hence that $H' = \phi(H) \subseteq A$ is a subgroup. Item (i) implies, moreover, that $\pi|_{H'} : H' \to H$ is an isomorphism.

If H is normal in $\langle\pi(A)\rangle$ then for every $a \in A$ and $h \in H$ there exists $\widehat{h} \in H$ such that $\pi(a^{-1})h\pi(a) = \widehat{h}$. In particular, $a^{-1}\phi(h)a\phi(\widehat{h}^{-1}) \in \ker\pi \cap A^4$, so if $A^4 \cap \ker\pi = \{1\}$ then $a^{-1}\phi(h)a \in H'$, and hence H' is normal in $\langle A\rangle$. □

Proof of Theorem 8.1.1 Let m be the quantity implied by the $O_K(1)$ notation in Proposition 8.1.3 (i), let ℓ be the word length of a simple commutator of weight $K^6 + 1$, and let $M \geq m(\ell + 1)$. Lemma 8.5.1 implies that there exists a homomorphism π from G to a nilpotent group N such that $A^{4M} \cap \ker\pi = \{1\}$. Applying Proposition 8.1.3 to $\pi(A)$, we conclude that there exist subgroups $H \lhd C \subseteq N$ such that $H \subseteq \pi(A^m)$, such that C/H is nilpotent of step at most K^6, such that C is generated by $\pi(A^6) \cap C$, and such that $|\pi(A^2) \cap C| \geq \exp(-K^{O(1)})|\pi(A)|$.

Define $C' = \langle A^m \cap \pi^{-1}(C)\rangle$, noting that $\pi(C') = C$. Note also that $H \subseteq \pi(A^m \cap \pi^{-1}(C)) = \pi(A^m \cap C')$, and so Lemma 8.5.2 implies that there is a normal subgroup $H' \lhd C'$ such that $H' \subseteq A^m$ and such that $\pi|_{H'} : H' \to H$ is an isomorphism.

Set $k = K^6$. Following [19], if $x_1, \ldots, x_{k+1} \in A^m \cap C'$ then the nilpotence of C/H implies that $[\pi(x_1), \ldots, \pi(x_{k+1})] \in H$, which implies that there exists $h \in H'$ such that $[\pi(x_1), \ldots, \pi(x_{k+1})]\pi(h) = 1$. By Lemma 8.5.2 (i), this implies that $[x_1, \ldots, x_{k+1}]h = 1$, and so we conclude that C'/H' is nilpotent of step at most K^6.

Finally, note that $\pi(A^2) \cap C = \pi(A^2 \cap C')$. Lemma 8.5.2 (i) therefore implies that $|A^2 \cap C'| = |\pi(A^2) \cap C| \geq \exp(-K^{O(1)})|\pi(A)| = \exp(-K^{O(1)})|A|$, and so the theorem follows from Lemma 7.1.5. □

Exercises

8.1 Let G be a finitely generated nilpotent group and let A be a finite K-approximate subgroup. Suppose there exists a central subgroup $Z < G$ such that $G = AZ$. Show that there exists a coset nilprogression P of rank at most $K^{O(1)}$ and step 1 such that $A \subset P \subset A^{K^{O(1)}}$.

8.2 Let G be a residually nilpotent group in which every element has order at most m. Suppose that $A \subset G$ is a finite K-approximate group. Show that A can be covered by K^{30K^6+2} left cosets of a nilpotent subgroup contained in $A^{(3m+2)K^6+2}$. *This generalises Theorem 2.4.1 to arbitrary residually nilpotent groups containing only elements of bounded order. Hint: First treat the case in which G is nilpotent; start by showing that the group D_{k+1} given by Proposition 8.3.1 is contained in $A^{(3r+2)K^6+2}$.*

8.3 Use Exercise 6.1 to show that the bound on the rank of the coset nilprogression P given by Theorem 8.1.2 can be improved from $\exp(\exp(K^{O(1)}))$ to $\exp(O(K^{12}))$.

8.4 Show that a finitely generated free group is residually nilpotent.

9
Soluble Approximate Subgroups of $GL_n(\mathbb{C})$

9.1 Introduction

In this chapter and the next we investigate approximate subgroups of *linear* groups. A linear group is simply a group that is isomorphic to some subgroup of $GL_n(\mathbb{K})$, with $\mathbb{K}$ a field. This has been an especially fruitful setting in which to study approxiate groups, leading as it does to many of the results on expanders that we described in the preface.

In this book we focus on the case $\mathbb{K} = \mathbb{C}$. It is actually the case where $\mathbb{K}$ is a finite field that is most important for applications to expansion. However, here we choose to work in the complex setting, where we can make greater progress than we could for more general $\mathbb{K}$ whilst keeping the exposition self-contained.

In the present chapter we restrict attention to *soluble* linear groups, which we define imminently; in the next chapter we explain how to deduce a result for arbitrary subgroups of $GL_n(\mathbb{C})$.

Definition 9.1.1 (derived series) Given a group G, the *derived series* $G = G^{(0)} > G^{(1)} > G^{(2)} > \cdots$ of G is defined recursively via $G^{(i)} = [G^{(i-1)}, G^{(i-1)}]$ for each $i \in \mathbb{N}$.

Definition 9.1.2 (soluble group) A group G is *soluble* (also called *solvable*) if there exists some d such that $G^{(d)} = \{1\}$. The smallest such d is called the *derived length* of G.

It easily follows from Proposition 5.2.7 and Corollary 5.2.9 that every s-step nilpotent group is soluble of derived length at most $\lfloor \log_2 s \rfloor + 1$.

For a non-nilpotent example, given $n \in \mathbb{N}$ the group

$$\text{Upp}_n(\mathbb{C}) = \left\{ \begin{pmatrix} x_{11} & \cdots & x_{1n} \\ & \ddots & \vdots \\ 0 & & x_{nn} \end{pmatrix} : \begin{array}{c} x_{ij} \in \mathbb{C} \\ x_{ii} \neq 0 \end{array} \right\}$$

of $n \times n$ upper-triangular complex matrices is easily verified to be soluble.

In the case of soluble complex linear groups, such as $\text{Upp}_n(\mathbb{C})$, we have the following effective version of Theorem 7.1.1.

Theorem 9.1.3 (Breuillard–Green [13, Theorem 1.4]) *Let $K \geq 2$. Let $G < GL_n(\mathbb{C})$ be a soluble subgroup, and suppose that $A \subset G$ is a finite K-approximate group. Then there is a nilpotent subgroup $N < G$ of step at most n such that A is contained in a union of at most $K^{O_n(1)}$ left cosets of N.*

Note that a version of Theorem 7.1.3 then follows easily from Theorem 9.1.3 and Proposition 7.1.4. In Exercise 9.3 we invite the reader to refine this version of Theorem 7.1.3.

Remarks The assumption $K \geq 2$ is for notational convenience, as described in Remark 6.1.2. However, unlike Theorem 6.1.1, the exact-group version of Theorem 9.1.3 corresponding to the cases $K < 2$ is not completely trivial. Nonetheless, we invite the reader to confirm in Exercise 9.1 that a finite soluble subgroup H of $GL_n(\mathbb{C})$ has an abelian subgroup of index at most $O_n(1)$. This can be seen as an exact-group version of Theorem 9.1.3. The well-known *Jordan's theorem* for finite linear groups states that the same statement still holds without the assumption that H is soluble.

Gill and Helfgott [28] have proved a result similar to Theorem 9.1.3 for soluble subgroups of $GL_n(\mathbb{F}_p)$. There are some similarities between the two arguments, but the Gill–Helfgott result uses more algebraic group theory, and it is largely to avoid the need for this theory that we restrict to the complex setting. Tao [65] has also proved a version of Theorem 9.1.3 for arbitrary soluble groups, although the conclusion is both qualitatively less explicit and quantitatively weaker than Theorem 9.1.3.

In proving Theorem 9.1.3 we use the fact that $\text{Upp}_n(\mathbb{C})$ is in some sense a universal example of a soluble complex linear group, as follows.

Theorem 9.1.4 (Mal'cev [44]) *Let $G < GL_n(\mathbb{C})$ be a soluble subgroup. Then G contains a normal subgroup U of index at most $O_n(1)$ that is conjugate to a subgroup of* $\text{Upp}_n(\mathbb{C})$.

We give a proof of Theorem 9.1.4 in Sections 9.A and 9.B, following the argument from Wehrfritz's book [75]. One reason for this is to keep this chapter self-contained, but an arguably more important reason is that by restricting attention to the complex numbers we are able to simplify the treatment quite considerably (Wehrfritz proves Theorem 9.1.4 and many of its precursory results in far greater generality). Nonetheless, the reader keen to concentrate on the theory specific to approximate groups, and prepared to accept Theorem 9.1.4 as 'classical', could reasonably skip the appendix.

Armed with Theorem 9.1.4, we can deduce Theorem 9.1.3 from the following result.

Proposition 9.1.5 (Breuillard–Green) *Let $K \geq 2$, and suppose that $A \subset \mathrm{Upp}_n(\mathbb{C})$ is a finite K-approximate group. Then there is a nilpotent subgroup $N < \mathrm{Upp}_n(\mathbb{C})$ of step at most n such that $|A^{O_n(1)} \cap N| \geq K^{-O_n(1)}|A|$.*

Proof of Theorem 9.1.3 from Proposition 9.1.5 It follows from Theorem 9.1.4 that G has a subgroup U of index $O_n(1)$ that is conjugate to $\mathrm{Upp}_n(\mathbb{C})$, and Lemma 2.6.2 then implies that

$$|A^2 \cap U| \gg_n |A|. \tag{9.1.1}$$

Proposition 2.6.5 implies that $A^2 \cap U$ is a K^3-approximate group. Proposition 9.1.5 therefore implies that there is a nilpotent subgroup $N < U$ of step at most n such that $|A^{O_n(1)} \cap N| \geq K^{-O_n(1)}|A^2 \cap U|$, and hence $|A^{O_n(1)} \cap N| \geq K^{-O_n(1)}|A|$ by (9.1.1). The theorem then follows from Lemma 7.1.5. □

We carry out the proof of Proposition 9.1.5 in the next two sections.

9.2 The Sum–Product Phenomenon over $\mathbb{C}$

In this section we introduce a phenomenon called the *sum–product* phenomenon, which governs how the operations of taking sum sets and product sets interact with one another. In $\mathbb{C}$, the sum–product phenomenon essentially refers to the fact that a finite subset of $\mathbb{C}$ cannot simultaneously have small additive doubling and small multiplicative doubling, as follows.

Theorem 9.2.1 (Solymosi's sum–product theorem over $\mathbb{C}$ [60]) *Let $U, V, W \subset \mathbb{C}$ be finite sets such that $U \neq \{0\}$ and $W \neq \{0\}$. Then*

$$|U + V||UW| \geq \frac{|U|^{3/2}|V|^{1/2}|W|^{1/2}}{56}.$$

In particular, an arbitrary finite set $A \subset \mathbb{C}$ satisfies

$$\max\{|A + A|, |AA|\} \geq \frac{|A|^{5/4}}{\sqrt{56}}.$$

Theorem 9.2.1 can be thought of as saying that $\mathbb{C}$ contains no finite 'approximate subfields'.

Remark Theorem 9.2.1 does not necessarily hold without the assumptions that $U \neq \{0\}$ and $W \neq \{0\}$. Indeed, if $U = \{0\}$ then $|U+V||UW| = |V|$, which is contrary to the conclusion of the theorem if W is much larger than V. On the other hand, if $U = V = [n]$ and $W = \{0\}$ then $|U + V||UW| = 2n - 1$ but $|U|^{3/2}|V|^{1/2}|W|^{1/2} = n^2$.

Since multiplication of matrices involves both addition and multiplication of entries, it is not entirely surprising that a result such as Theorem 9.2.1 should feature in the proof of Theorem 9.1.3. Indeed, prior to the proof of Theorem 9.1.3 by Breuillard and Green, Helfgott [39, 40] had used sum–product results over finite fields when studying approximate subgroups of $SL_2(\mathbb{F}_p)$ and $SL_3(\mathbb{F}_p)$. On the other hand, in generalising Helfgott's work to higher dimensions, Breuillard, Green and Tao [16] do not use any sum–product results; in fact, they recover a sum–product result over $\mathbb{F}_p$ from their work [16, Theorem 2.7].

We refer the reader to [63] for a general introduction to the sum–product phenomenon.

The proof of Theorem 9.2.1 relies on the geometry of the complex plane, via the following results.

Lemma 9.2.2 *Let a, b, c be the lengths of the sides of a Euclidean triangle, and write α, β, γ, respectively, for the interior angles opposite those sides. Suppose that $\alpha \leq \beta \leq \gamma$. Then $a \leq b \leq c$.*

Proof By the sine rule it suffices to show that $\sin\alpha \leq \sin\beta \leq \sin\gamma$. If $\gamma \leq \pi/2$ this is immediate, since sin is increasing on $[0, \pi/2]$. On the other hand, if $\gamma > \pi/2$ then we have $\alpha \leq \beta < \pi - \gamma < \pi/2$, and so $\sin\alpha \leq \sin\beta < \sin(\pi - \gamma) = \sin\gamma$. □

In the next two lemmas, given $z \in \mathbb{C}$ and $r > 0$ we write $D(z, r) = \{\omega \in \mathbb{C} : |z - \omega| < r\}$, the open disc of radius r centred at $z \in \mathbb{C}$, and $\overline{D}(z, r) = \{\omega \in \mathbb{C} : |z - \omega| \leq r\}$, the closed disc.

Lemma 9.2.3 *Suppose that $z_1, \ldots, z_k \in \mathbb{C}$ and $r_1, \ldots, r_k > 0$ are such that the discs $D(z_i, r_i)$ satisfy*

$$z_i \notin D(z_j, r_j) \qquad (i \neq j) \tag{9.2.1}$$

and

$$\bigcap_{i=1}^{k} \overline{D}(z_i, r_i) \neq \varnothing. \tag{9.2.2}$$

Then $k \leq 7$.

Proof By (9.2.2) we may fix $x \in \bigcap_{i=1}^{k} \overline{D}(z_i, r_i)$. We first prove the lemma in the special case where $z_i \neq x$ for every i. In that case, for notational convenience we may relabel the z_i so that $\arg(z_i - x)$ is non-decreasing in i. We then consider the suffices of the z_i as integers modulo k, so that $z_{k+1} = z_1$. The assumption (9.2.1) then implies that $|z_i - z_{i+1}| \geq |z_i - x|, |z_{i+1} - x|$ for every i, and so Lemma 9.2.2 implies that $\arg(z_{i+1} - x) - \arg(z_i - x) \geq \pi/3$ for every i. It follows that $k \leq 6$.

If $z_i = x$ for some i then removing z_i and r_i from the lists under consideration decreases k by at most 1 without violating the hypothesis of the lemma, and so the special case of the previous paragraph implies that $k - 1 \leq 6$, and hence $k \leq 7$, as required. □

Lemma 9.2.4 *Let $U, V, W \subset \mathbb{C}$ be finite sets and suppose that $0 \notin W$ and $|U| \geq 2$. Fix $v \in V$ and $w \in W$, and for each $u \in U$ fix an element $n(u) \in U \setminus \{u\}$ that minimises $|u - n(u)|$ (thus $n(u)$ is the 'nearest neighbour' to u in U). Then*

$$\sum_{u \in U} \big|\{x \in U + V : |(u + v) - x| \leq |u - n(u)|\}\big| \leq 7|U + V| \tag{9.2.3}$$

and

$$\sum_{u \in U} \big|\{x \in UW : |uw - x| \leq |uw - n(u)w|\}\big| \leq 7|UW|. \tag{9.2.4}$$

Proof For each $u \in U$ we have $D(u, |u - n(u)|) \cap U = \{u\}$, and so Lemma 9.2.3 implies that any given $x \in \mathbb{C}$ satisfies $x \in D(u, |u - n(u)|)$ for at most seven distinct $u \in U$. Translating by v or dilating by w, this means that every point $x \in \mathbb{C}$ satisfies $x \in D(u + v, |u - n(u)|)$ for at most seven distinct $u \in U$, and satisfies $x \in D(uw, w|u - n(u)|)$ for at most seven distinct $u \in U$. This means in particular that each $x \in U+V$ contributes 1 to at most seven of the summands in (9.2.3), whilst each $x \in UW$ contributes 1 to at most seven of the summands in (9.2.4), and the lemma is proved. □

Proof of Theorem 9.2.1 If $|U| = 1$ then $|U+V||UW| = |V||W|$ and the theorem holds, so we may assume that $|U| \geq 2$. Set $W_0 = W \setminus \{0\}$, noting that

$$\tfrac{1}{2}|W| \leq |W_0| \leq |W|. \tag{9.2.5}$$

For each $u \in U$ fix an element $n(u) \in U \setminus \{u\}$ that minimises $|u - n(u)|$, as in the hypothesis of Lemma 9.2.4.

We first note that for each $v \in V$ at least three quarters of the elements $u \in U$ satisfy

$$\big|\{x \in U+V : |(u+v) - x| \leq |u - n(u)|\}\big| \leq \frac{28|U+V|}{|U|}, \tag{9.2.6}$$

since if more than a quarter of them violated this inequality then their contributions to the sum (9.2.3) would total more than $6|U+V|$ and contradict Lemma 9.2.4. Similarly, for each $w \in W_0$ at least three quarters of the elements $u \in U$ satisfy

$$\big|\{x \in UW_0 : |uw - x| \leq |uw - n(u)w|\}\big| \leq \frac{28|UW|}{|U|}. \tag{9.2.7}$$

In particular, there are at least $\frac{1}{2}|U||V||W_0|$ triples $(u, v, w) \in U \times V \times W_0$ satisfying both (9.2.6) and (9.2.7). On the other hand, every such triple is uniquely determined by the quadruple $(u+v, n(u)+v, uw, n(u)w)$, since we can recover w by noting that

$$w = \frac{uw - n(u)w}{(u+v) - (n(u)+v)},$$

and then, since $w \neq 0$, we can immediately recover u and v. For such a quadruple there are at most $|U+V|$ possibilities for the first element, and then by (9.2.6) the second element must be one of the $28|U+V|/|U|$ elements of $U+V$ that are closest to the first element. Similarly, there are at most $|UW_0|$ possibilities for the third element, and then the fourth element must be one of the $28|UW_0|/|U|$ elements of UW_0 that are closest to the third element. In particular, there are at most

$$\frac{28^2|U+V|^2|UW_0|^2}{|U|^2}$$

quadruples $(u+v, u'+v, uw, u'w)$ with $u \in U$, $v \in V$ and $w \in W_0$ satisfying both (9.2.6) and (9.2.7), and hence at most that number of triples $(u, v, w) \in U \times V \times W_0$ satisfying both (9.2.6) and (9.2.7). Comparing this to the lower bound we had previously, this implies that

$$\frac{28^2|U+V|^2|UW_0|^2}{|U|^2} \geq \frac{|U||V||W_0|}{2},$$

and hence

$$\begin{aligned} |U+V||UW| &\geq |U+V||UW_0| \\ &\geq \frac{|U|^{3/2}|V|^{1/2}|W_0|^{1/2}}{28\sqrt{2}} \\ &\geq \frac{|U|^{3/2}|V|^{1/2}|W|^{1/2}}{56} \end{aligned}$$

by (9.2.5), as required. □

9.3 Complex Upper-Triangular Groups

In this section we complete the proof of Proposition 9.1.5, essentially following Breuillard and Green [13].

Proof of Proposition 9.1.5 The case $n = 1$ is trivial because $\mathrm{Upp}_1(\mathbb{C}) \cong \mathbb{C}^\times$ is abelian, so we may assume $n \geq 2$ and proceed by induction. To perform the inductive step we define two homomorphisms $\pi_1, \pi_2 : \mathrm{Upp}_n(\mathbb{C}) \to \mathrm{Upp}_{n-1}(\mathbb{C})$ projecting to the following submatrices:

$$\begin{pmatrix} x_{11} & x_{12} & \cdots & x_{1(n-1)} & x_{1n} \\ 0 & x_{22} & \cdots & x_{2(n-1)} & x_{2n} \\ \vdots & \vdots & \ddots & \vdots & \vdots \\ 0 & 0 & \cdots & x_{(n-1)(n-1)} & x_{(n-1)n} \\ 0 & 0 & \cdots & 0 & x_{nn} \end{pmatrix}.$$

Thus

$$\pi_1 \begin{pmatrix} x_{11} & x_{12} & \cdots & x_{1(n-1)} & x_{1n} \\ & x_{22} & \cdots & x_{2(n-1)} & x_{2n} \\ & & \ddots & \vdots & \vdots \\ & & & x_{(n-1)(n-1)} & x_{(n-1)n} \\ & & & & x_{nn} \end{pmatrix} = \begin{pmatrix} x_{11} & x_{12} & \cdots & x_{1(n-1)} \\ & x_{22} & \cdots & x_{2(n-1)} \\ & & \ddots & \vdots \\ & & & x_{(n-1)(n-1)} \end{pmatrix}$$

and

$$\pi_2 \begin{pmatrix} x_{11} & x_{12} & \cdots & x_{1(n-1)} & x_{1n} \\ & x_{22} & \cdots & x_{2(n-1)} & x_{2n} \\ & & \ddots & \vdots & \vdots \\ & & & x_{(n-1)(n-1)} & x_{(n-1)n} \\ & & & & x_{nn} \end{pmatrix} = \begin{pmatrix} x_{22} & \cdots & x_{2(n-1)} & x_{2n} \\ & \ddots & \vdots & \vdots \\ & & x_{(n-1)(n-1)} & x_{(n-1)n} \\ & & & x_{nn} \end{pmatrix}.$$

Since $\pi(A)$ is a K-approximate group, by the induction hypothesis there is a nilpotent subgroup $N_1 < \mathrm{Upp}_{n-1}(\mathbb{C})$ of step at most $n-1$ and some $k = k_n$ such that $|\pi_1(A)^k \cap N_1| \geq K^{-O_n(1)}|\pi_1(A)|$. Lemma 6.2.2 then implies that

$$|A^{k+2} \cap \pi_1^{-1}(N_1)| \geq K^{-O_n(1)}|A|. \tag{9.3.1}$$

The set $A^{k+2} \cap \pi_1^{-1}(N_1)$ is a $K^{O_n(1)}$-group by Proposition 2.6.5, and hence so is $\pi_2(A^{k+2} \cap \pi_1^{-1}(N_1))$. Applying the induction hypothesis again we conclude that there exists a nilpotent subgroup $N_2 < \mathrm{Upp}_{n-1}(\mathbb{C})$ of step at most $n-1$ such that

$$\left|\pi_2\left(A^{k^2+2k} \cap \pi_1^{-1}(N_1)\right) \cap N_2\right| \geq K^{-O_n(1)}\left|\pi_2\left(A^{k+2} \cap \pi_1^{-1}(N_1)\right)\right|.$$

Applying Lemma 6.2.2 again then implies that

$$\begin{aligned} |A^{k^2+2k+2} \cap \pi_1^{-1}(N_1) \cap \pi_2^{-1}(N_2)| &\geq K^{-O_n(1)}|A^{k+2} \cap \pi_1^{-1}(N_1)| \\ &\geq K^{-O_n(1)}|A| \end{aligned}$$

by (9.3.1). Writing $B = A^{k^2+2k+2} \cap \pi_1^{-1}(N_1) \cap \pi_2^{-1}(N_2)$, we therefore have

$$B \subset A^{O_n(1)} \tag{9.3.2}$$

and

$$|B| \geq K^{-O_n(1)}|A|. \tag{9.3.3}$$

Moreover, B is a $K^{O_n(1)}$-group by Proposition 2.6.5.

Since $B \subset \pi_1^{-1}(N_1) \cap \pi_2^{-1}(N_2)$ and both N_1 and N_2 are $(n-1)$-step nilpotent we have $[\langle B\rangle, \ldots, \langle B\rangle]_n \subset \ker \pi_1 \cap \ker \pi_2$ by Corollary 5.2.9. Given $\lambda \in \mathbb{C}$, write $m_\lambda \in \mathrm{Upp}_n(\mathbb{C})$ for the matrix

$$m_\lambda = \begin{pmatrix} 1 & 0 & \ldots & 0 & \lambda \\ 0 & 1 & \ldots & 0 & 0 \\ \vdots & \vdots & \ddots & \vdots & \vdots \\ 0 & 0 & \ldots & 1 & 0 \\ 0 & 0 & \ldots & 0 & 1 \end{pmatrix}.$$

Note that

$$m_\mu m_\lambda = m_{\mu+\lambda} \tag{9.3.4}$$

for every $\mu, \lambda \in \mathbb{C}$, and hence that the set $Z = \{m_\lambda : \lambda \in \mathbb{C}\}$ forms a subgroup. Alternatively, this follows from the observation that $Z = \ker \pi_1 \cap \ker \pi_2$, which also implies that

$$[\langle B\rangle, \ldots, \langle B\rangle]_n \subset Z. \tag{9.3.5}$$

Writing M for the length of the commutator $[b_1, \ldots, b_n]$ as a word in the b_i, we therefore have

$$[B, \ldots, B]_n \subset B^M \cap Z. \tag{9.3.6}$$

Writing $S = \{\lambda \in \mathbb{C} : m_\lambda \in B^M \cap Z\}$, it follows from (9.3.4) that

$$B^{2M} \cap Z \supset \{m_\lambda : \lambda \in S + S\}. \tag{9.3.7}$$

Defining a homomorphism $\tau : \mathrm{Upp}_n(\mathbb{C}) \to \mathbb{C}^\times$ via

$$\tau \begin{pmatrix} x_{11} & \ldots & x_{1n} \\ \vdots & \ddots & \vdots \\ 0 & \ldots & x_{nn} \end{pmatrix} = \frac{x_{11}}{x_{nn}},$$

we have

$$x m_\lambda x^{-1} = m_{\tau(x)\cdot\lambda} \tag{9.3.8}$$

for every $x \in \mathrm{Upp}_n(\mathbb{C})$ and $\lambda \in \mathbb{C}$, and hence

$$B^{M+2} \cap Z \supset \{m_\lambda : \lambda \in \tau(B) \cdot S\}. \tag{9.3.9}$$

Combining (9.3.7) and (9.3.9) (and noting that $M \geq 2$, since $n \geq 2$), we conclude that

$$|B^{2M} \cap Z| \geq \max\{|S+S|, |\tau(B) \cdot S|\}. \tag{9.3.10}$$

We now claim that there exists a nilpotent group N of step at most n such that

$$|B^2 \cap N| \geq K^{-O_n(1)}|B|. \tag{9.3.11}$$

If $S = \{0\}$ then $\langle B \rangle$ is nilpotent of step at most $n-1$ by (9.3.6) and Lemma 5.2.3, and the claim is trivial. We may therefore assume that $S \neq \{0\}$. Since $0 \notin \tau(\mathrm{Upp}_n(\mathbb{C}))$ by definition, we also have $\tau(B) \neq \{0\}$. We may therefore apply Theorem 9.2.1 with $U = V = S$ and $W = \tau(B)$ to conclude that $|S+S||\tau(B) \cdot S| \gg |S|^2|\tau(B)|^{1/2}$, and hence that $\max\{|S+S|, |\tau(B) \cdot S|\} \gg |S||\tau(B)|^{1/4}$. Since $|S| = |B^M \cap Z|$ by definition, this combines with (9.3.10) to imply that

$$|B^{2M} \cap Z| \gg |\tau(B)|^{1/4}|B^M \cap Z|.$$

Since B is a $K^{O_n(1)}$-approximate group, Proposition 2.6.5 therefore implies that $|\tau(B)| \leq K^{O_n(1)}$, and so Lemma 2.6.2 implies that

$$|B^2 \cap \ker \tau| \geq K^{-O_n(1)}|B|. \tag{9.3.12}$$

It follows from (9.3.8) that $\ker \tau$ commutes with Z, and so (9.3.5) implies that for every $x \in \ker \tau$ and $b_1, \ldots, b_n \in B^2$ we have $[b_1, \ldots, b_n, x] = 1$. In particular, for every $b_1, \ldots, b_{n+1} \in B^2 \cap \ker \tau$ we have $[b_1, \ldots, b_{n+1}] = 1$, and so by Lemma 5.2.3 $B^2 \cap \ker \tau$ generates an n-step nilpotent group, say N. It then follows from (9.3.12) that N satisfies (9.3.11) as claimed.

The proposition then follows from (9.3.2) and (9.3.3). □

9.A Representation Theory

In this appendix we give a self-contained proof of Theorem 9.1.4, following Wehrfritz [75] but simplifying matters by restricting certain results to the complex field. The proof uses a certain amount of representation theory, and in this section we introduce that theory; in the next section we apply it to prove Theorem 9.1.4.

We start with some basic terminology.

Definition 9.A.1 (representation) Given a group G and a vector space V, a *representation* of G on V is a homomorphism $\rho : G \to GL(V)$. The dimension of V is called the *degree* of the representation. A representation ρ is said to be *faithful* if $\ker \rho = \{1\}$. If U is a vector subspace of V that is invariant under G in the sense that $\rho(G)(U) = U$ then the representation ρ_U defined by $\rho_U(g) = \rho(g)|_U$ is said to be a *subrepresentation* of ρ. If the only subrepresentation of ρ is ρ itself then ρ is said to be *irreducible.* If V can be decomposed as a direct sum of G-invariant subspaces then ρ is said to be *completely reducible*, and its irreducible subrepresentations are called its *irreducible components.*

Definition 9.A.2 (homomorphism of representations) Let V, W be vector spaces over the same field, and suppose $\rho_1 : G \to GL(V)$ and $\rho_2 : G \to GL(W)$ are two representations of the same group G. A linear map $\varphi : V \to W$ is said to be a *homomorphism of representations*, or just a *homomorphism*, from ρ_1 to ρ_2 if

$$\varphi \circ \rho_1(g) = \rho_2(g) \circ \varphi$$

for every $g \in G$. If φ is bijective then it is said to be an *isomorphism.* We write $\mathrm{Hom}(\rho_1, \rho_2)$ for the space of all such homomorphisms of representations.

Note that a group is linear if and only if it admits a faithful representation. If G is actually a subgroup of $GL(V)$ then we often say that G itself is 'completely reducible' or 'irreducible' to mean that the inclusion representation

$$\begin{aligned} \iota_G : G &\hookrightarrow GL(V) \\ g &\mapsto g \end{aligned}$$

is completely reducible or irreducible. For a given representation ρ of G we also occasionally say that G 'acts irreducibly via ρ' to mean that ρ is irreducible; if it is clear from the context what ρ is then we sometimes drop the 'via ρ'.

In the proof of Theorem 9.1.4 we need only consider *complex representations*, which is to say representations on complex vector spaces, and so in much of what we present here we restrict to that case. The reader can find fully general versions of this material in Wehrfritz [75].

Breaking a representation into irreducible components can be thought of as similar in spirit to diagonalising a matrix. Indeed, it is a simple exercise to check that complete reducibility of a representation $\rho : G \to$

$GL_n(\mathbb{K})$ is equivalent to the existence of irreducible representations $\rho_i : G \to GL_{n_i}(\mathbb{K})$ for some $n_1, \ldots, n_r$ with $n_1 + \cdots + n_r = n$ and a linear map $\alpha \in GL_n(\mathbb{K})$ such that

$$\alpha^{-1}\rho(g)\alpha = \begin{pmatrix} \rho_1(g) & & 0 \\ & \ddots & \\ 0 & & \rho_r(g) \end{pmatrix} \tag{9.A.1}$$

for every $g \in G$. Of course, not all matrices are diagonalisable, and it turns out that neither are all representations completely reducible. Indeed, the former statement gives an easy way to verify the latter, since it is not hard to check that if the matrix $M \in GL_n(\mathbb{C})$ is not diagonalisable then the representation $\rho : \mathbb{Z} \to GL_n(\mathbb{C})$ defined by setting $\rho(n) = M^n$ is not completely reducible.

Nonetheless, it is well known and often very convenient that even if a complex matrix is not diagonalisable it is always upper-triangularisable. The following simple lemma can be thought of as an analogue of this fact: it shows that even if a representation is not completely reducible – that is to say, not 'diagonalisable' in the sense of (9.A.1) – it is at least 'upper-triangularisable' in the following sense.

Lemma 9.A.3 *Let G be a group, let $\mathbb{K}$ be a field and let $\rho : G \to GL_n(\mathbb{K})$ be a representation of G. Then there exist $\alpha \in GL_n(\mathbb{K})$ and irreducible representations $\rho_i : G \to GL_{n_i}(\mathbb{K})$ for some $n_1, \ldots, n_r$ with $n_1 + \cdots + n_r = n$ such that*

$$\alpha^{-1}\rho(g)\alpha = \begin{pmatrix} \rho_1(g) & & * \\ & \ddots & \\ 0 & & \rho_r(g) \end{pmatrix}$$

for every $g \in G$.

Proof Write $e_1, \ldots, e_n$ for the standard basis of $\mathbb{K}^n$. If ρ is irreducible then the lemma holds with $r = 1$, so we may assume that there is a G-invariant subspace U of dimension n_1 with $0 < n_1 < n$. Fix a basis $b_1, \ldots, b_{n_1}$, and define $\rho_1 : G \to GL_{n_1}(\mathbb{K})$ by mapping each g to the matrix that represents $\rho(g)|_U$ with respect to this basis. Extending $b_1, \ldots, b_{n_1}$ to a basis $b_1, \ldots, b_n$ for $GL_n(\mathbb{K})$, and defining α_0 to be the linear map taking each $e_i \mapsto b_i$, we have

$$\alpha_0^{-1}\rho(g)\alpha_0 = \begin{pmatrix} \rho_1(g) & * \\ 0 & \theta(g) \end{pmatrix}$$

for some map $\theta : G \to GL_{n-n_1}(\mathbb{K})$. However, it is easy to verify that θ is a homomorphism, and hence a representation of G of degree $n-n_1 < n$, and so the lemma follows by induction on n. □

There is therefore an extent to which even representations that are not completely reducible can nonetheless be understood in terms of irreducible representations. Indeed, this approach will be critical in our eventual proof of Theorem 9.1.4. With this in mind, we spend the next few paragraphs introducing some of the basic theory of irreducible representations, starting with one of the fundamental results of representation theory: Schur's lemma.

Lemma 9.A.4 (Schur) *Let G be a group, let V, W be vector spaces over the same field $\mathbb{K}$, and let $\rho : G \to GL(V)$ and $\pi : G \to GL(W)$ be irreducible representations of G. Then*

(i) every homomorphism of representations from ρ to π is either 0 or an isomorphism; and
(ii) if $\mathbb{K}$ is algebraically closed (in particular, if $\mathbb{K} = \mathbb{C}$) then every homomorphism of representations from ρ to itself is a scalar multiple of the identity.

Proof

(i) If φ is a homomorphism from ρ to π then $\ker\varphi$ is a subrepresentation of ρ and $\varphi(V)$ is a subrepresentation of π. The irreducibility of ρ and π therefore implies that either $\ker\varphi = \{0\}$ and $\varphi(V) = W$, in which case φ is an isomorphism, or $\varphi(V) = \{0\}$, in which case $\varphi = 0$.
(ii) Since $\mathbb{K}$ is algebraically closed, $\varphi : V \to V$ has an eigenvalue λ. It follows that $\varphi - \lambda \cdot 1$ is not invertible, and so must be 0 by part (i).

□

Given a complex vector space V, we use the notation $\mathbb{C} \cdot 1$ to mean the subgroup of $GL(V)$ consisting of scalar multiples of the identity.

Corollary 9.A.5 *Let G be a group and let $\rho : G \to GL_n(\mathbb{C})$ be an irreducible representation. Then $\rho(Z(G)) \subset \mathbb{C} \cdot 1$.*

Proof Given an arbitrary $z \in Z(G)$ we have $\rho(z) \circ \rho(g) = \rho(g)\rho(z)$ for every $g \in G$ by definition of the centre. It follows that $\rho(z)$ is a homomorphism of representations from ρ to itself, and so it is a scalar multiple of the identity by Lemma 9.A.4 (ii). □

Corollary 9.A.6 *Every irreducible complex representation of a finite abelian group has dimension* 1.

Note therefore that the irreducible representations of a finite abelian group G are precisely the Fourier characters $\zeta_\gamma : G \to \mathbb{C}^\times$ defined in Section 4.2.

The next lemma is trivial, but we record it for ease of later reference.

Lemma 9.A.7 *Let G be a group, let $\rho : G \to GL(V)$ be a representation of G, and let U, W be invariant subspaces of V for ρ. Then $U \cap W$ is also an invariant subspace. In particular, if U and W are irreducible then either $U = W$ or $U \cap W = \{0\}$.*

The next result, which is one part of a more extensive theorem of Clifford, looks at what happens when we restrict an irreducible representation to a normal subgroup.

Lemma 9.A.8 (Clifford) *Let G be a group, let $\rho : G \to GL(V)$ be an irreducible representation, and let $H \lhd G$ be a normal subgroup of G. Then the restriction $\rho|_H : H \to GL(V)$ is completely reducible, and all of its irreducible subrepresentations have the same kernel in H.*

Proof Let U be an irreducible subspace of V for $\rho|_H$, noting that $\sum_{g\in G} \rho(g)(U)$ is an invariant subspace of V for ρ, and hence that

$$\sum_{g\in G} \rho(g)(U) = V \tag{9.A.2}$$

by the irreducibility of V. The invariance of U for $\rho|_H$ and the normality of H imply that for every $g \in G$ and $h \in H$ we have

$$\begin{aligned}\rho(h)\rho(g)U &= \rho(g)\rho(h^g)U \\ &\subset \rho(g)U,\end{aligned}$$

and so $\rho(g)U$ is also irreducible for $\rho|_H$. It therefore follows from (9.A.2) that V is spanned by subspaces that are irreducible for $\rho|_H$, and so the last part of Lemma 9.A.7 implies that $\rho|_H$ is a direct sum of finitely many irreducible representations

$$\begin{aligned}\rho_g : H &\to \rho(g)(U) \\ h &\mapsto \rho(h)|_{\rho(g)(U)},\end{aligned}$$

as required. To see that these all have the same kernel, note that for

$u \in U$, $h \in H$ and $g \in G$ we have $\rho(h)\rho(g)u = \rho(g)\rho(h^g)u$, so that

$$\begin{aligned} h \in \ker(\rho_g) &\iff h^g \in \ker(\rho_1) \\ &\iff h \in \ker(\rho_1). \end{aligned}$$

□

We now move on to discuss another way of decomposing representations into direct sums of smaller subspaces: *systems of imprimitivity.*

Definition 9.A.9 (system of imprimitivity) Given a representation $\rho : G \to GL(V)$ of a group G in a vector space V, a *system of imprimitivity* for ρ is a set $\{V_1, \ldots, V_r\}$ of non-trivial subspaces of V such that $V = V_1 \oplus \cdots \oplus V_r$ and such that for every $i \in [r]$ and every $g \in G$ there exists j such $\rho(g)(V_i) = V_j$. In this set-up, the singleton $\{V\}$ consisting only of the space V itself is always a system of imprimitivity for ρ. If there are no other systems of imprimitivity for ρ then ρ is said to be *primitive*; otherwise, ρ is said to be *imprimitive*.

As with irreducibility, if G is a subgroup of $GL(V)$ then we often say that $\{V_1, \ldots, V_r\}$ is a system of imprimitivity for G if it is a system of imprimitivity for the inclusion representation $\iota_G : G \hookrightarrow GL(V)$; we also often say that G is primitive or imprimitive if ι_G is. Likewise, for a given representation ρ of G, we occasionally say that G 'acts primitively via ρ' to mean that ρ is primitive, and if it is clear from the context what ρ is then we often drop the 'via ρ'.

Lemma 9.A.10 *Let G be a group and let $\rho : G \to GL(V)$ be an irreducible representation. Suppose that $\{V_1, \ldots, V_r\}$ is a system of imprimitivity for ρ. Then $\rho(G)$ acts transitively on $\{V_1, \ldots, V_r\}$.*

Proof The space $\sum_{g \in G} \rho(g)(V_1)$ is a non-zero invariant subspace of V for ρ, so is equal to V by irreducibility. □

Given a representation $\rho : G \to GL(V)$ of a group G and a subspace $W < V$, we define

$$N_G(W) = \{g \in G : \rho(g)(W) = W\}$$

and

$$C_G(W) = \{g \in G : \rho(g)(w) = w \text{ for every } w \in W\}.$$

If $G < GL(V)$ then unless otherwise stated we take ρ to be the inclusion representation in these definitions.

Lemma 9.A.11 *Let G be a subgroup of $GL(V)$ and let $V_1, \ldots, V_r$ be a system of imprimitivity for G. Then $N = \bigcap_{i=1}^r N_G(V_i)$ is a normal subgroup of G and satisfies $[G : N] \leq r!$.*

Proof The action of G on the subspaces V_i defines a homomorphism of G into the symmetric group on $[r]$, of which N is the kernel. □

A system of imprimitivity $\{V_1, \ldots, V_r\}$ for a representation ρ is said to be *minimal* if $N_\rho(V_i)$ acts primitively on V_i via ρ for every i.

Lemma 9.A.12 *Let G be a group, and let $\rho : G \to GL(V)$ be an irreducible representation. Suppose that $\{V_1, \ldots, V_r\}$ is a system of imprimitivity for G in V, and for each $i, j \in [r]$ let $g_{i,j} \in G$ be such that $\rho(g_{i,j})(V_i) = V_j$ (the elements $g_{i,j}$ exist by Lemma 9.A.10). Then the following hold.*

(i) The restriction of ρ to $N_G(V_i)$ is irreducible on V_i.
(ii) If $\{W_1, \ldots, W_t\}$ is a system of imprimitivity for $N_G(V_i)$ in V_i then $\{\rho(g_{i,j})(W_\ell) : j \in [r], \ell \in [t]\}$ is a system of imprimitivity for G in V. In particular, if ρ is of finite degree then G has a minimal system of imprimitivity in V.

Proof First note that given $g \in G$ and $i, j \in [r]$ there exists $k = k(g, i, j)$ such that $\rho(gg_{i,j})(V_i) = V_k$ by definition of a system of imprimitivity, and hence

$$\rho(g_{i,k}^{-1} g g_{i,j}) \in N_G(V_i). \tag{9.A.3}$$

We now prove the two conclusions of the lemma.

(i) We need to show that an arbitrary non-zero $N_G(V_i)$-invariant subspace of $U_i < V_i$ is equal to V_i. Given $g \in G$ and $j \in [r]$, (9.A.3) implies in particular that $\rho(g_{i,k}^{-1} g g_{i,j})(U_i) \subset U_i$, and hence that $\rho(gg_{i,j})(U_i) \subset \rho(g_{i,k})(U_i)$. It follows that $\sum_{j=1}^r \rho(g_{i,j})(U_i)$ is a non-zero G-invariant subspace of V, and hence equal to V, which is only possible if $U_i = V_i$, as required.
(ii) Given $j \in [r]$ and $\ell \in [t]$, (9.A.3) implies that for $g \in G$ there exists m such that $\rho(g_{i,k}^{-1} g g_{i,j})(W_\ell) = W_m$, and hence $\rho(g)(\rho(g_{i,j})(W_\ell)) = \rho(g_{i,k})(W_m)$.

□

Lemma 9.A.13 *Let G be an irreducible subgroup of $GL_n(\mathbb{C})$, and suppose that A is an abelian normal subgroup of G. Then A is simultaneously* *diagonalisable and G permutes the eigenspaces of A. In particular, the eigenspaces of A form a system of imprimitivity for G.*

Proof Since A is abelian, all of its irreducible subrepresentations have dimension 1 by Corollary 9.A.6. Lemma 9.A.8 implies that A is completely reducible, and so A is simultaneously diagonalisable, as required.

Now suppose that V is an eigenspace for A, so that there exists a function $\lambda : A \to \mathbb{C}$ such that $a(V) = \lambda(a)V$ for every $a \in A$, and let $g \in G$. It remains to show that $g(V)$ is also an eigenspace for A. However, this follows from the normality of A, since for every $a \in A$ we have $ag(V) = g(\lambda(a^g)V) = \lambda(a^g)g(V)$. □

Lemma 9.A.13 immediately implies the following result.

Corollary 9.A.14 (Blichfeldt) *Let G be a primitive irreducible subgroup of $GL_n(\mathbb{C})$, and suppose that A is an abelian normal subgroup of G. Then $A \subset \mathbb{C} \cdot 1$.*

Corollary 9.A.15 *Let G be a primitive irreducible subgroup of $GL_n(\mathbb{C})$. Then $Z(G)$ is the unique maximal abelian normal subgroup of G.*

9.B The Structure of Soluble Linear Groups

In this section we prove Theorem 9.1.4, following Wehrfritz's book [75]. The main ingredient is the following proposition.

Proposition 9.B.1 *Let G be an irreducible soluble subgroup of $GL_n(\mathbb{C})$. Then G contains a normal abelian subgroup of index $n!((n^2)!n^2)^n$.*

The reader should not attach too much importance to the precise form of the bound $n!((n^2)!n^2)^n$, which we make no attempt to optimise. The key point is that it depends only on n.

We first prove Proposition 9.B.1 in the primitive case, as follows.

Proposition 9.B.2 *Let G be a primitive irreducible soluble subgroup of $GL_n(\mathbb{C})$ and let A be a normal subgroup of G that is maximal with respect to the condition that $A/Z(G)$ is abelian. Then $[A : Z(G)] \leq n^2$ and $[G : A] \leq (n^2)!$. In particular, $[G : Z(G)] \leq (n^2)!n^2$.*

The group A in Proposition 9.B.2 is automatically 2-step nilpotent, so the following lemma is a step in the right direction.

Lemma 9.B.3 *Let G be an irreducible 2-step nilpotent subgroup of $GL_n(\mathbb{C})$, and suppose that $g_1, \ldots, g_r$ belong to distinct cosets of $Z(G)$. Then the matrices g_i are linearly independent over $\mathbb{C}$. In particular, $[G : Z(G)] \leq n^2$.*

Proof Suppose that the matrices $g_1, \ldots, g_r$ are not linearly independent over $\mathbb{C}$. Reordering if necessary, we may therefore assume that there is a linear dependency

$$\lambda_1 g_1 + \cdots + \lambda_s g_s = 0 \tag{9.B.1}$$

with the minimum number of terms. In particular, each $\lambda_i \neq 0$ and $s \geq 2$.

Since $g_1^{-1} g_2 \notin Z(G)$ there exists $x \in G$ such that $[g_1^{-1} g_2, x] \neq 1$. Lemma 5.5.2 and the nilpotence of G therefore imply that $[g_1, x] \neq [g_2, x]$. The nilpotence of G implies that $[g_i, x] \in Z(G)$ for each i, and so Corollary 9.A.5 implies that for each i there exists $\alpha_i \in \mathbb{C}$ such that $[g_i, x] = \alpha_i \cdot 1$, with $\alpha_1 \neq \alpha_2$ by the definition of x. Rewriting $[g_i, x]$ as $g_i^{-1} g_i^x$, we have $g_i^{-1} g_i^x = \alpha_i \cdot 1$, and hence

$$g_i^x = \alpha_i g_i \tag{9.B.2}$$

for each i.

Now (9.B.1) implies in particular that

$$\alpha_1 \sum_{i=1}^{s} \lambda_i g_i = 0$$

and

$$\sum_{i=1}^{s} \lambda_i g_i^x = 0,$$

which combined with (9.B.2) implies that

$$\sum_{i=2}^{s} (\alpha_1 - \alpha_i) \lambda_i g_i = 0.$$

Since $\alpha_1 \neq \alpha_2$ and $\lambda_2 \neq 0$, this contradicts the minimality of the number of terms in (9.B.1). It follows that the matrices $g_1, \ldots, g_r$ are indeed linearly independent, as required. □

Proof of Proposition 9.B.2 Abbreviate $Z = Z(G)$. Since $Z(A)$ is characteristic in A, Lemma 1.5.2 implies that it is normal in G. Since it is also abelian, Corollary 9.A.14 therefore implies that $Z(A) \subset Z$. Since $Z \subset Z(A)$, we conclude that

$$Z(A) = Z. \tag{9.B.3}$$

Now it follows from Lemma 9.A.8 that A is completely reducible into irreducible subrepresentations $\rho_1, \ldots, \rho_r$, say, all of whose kernels are equal. Spelling out the equality of their kernels explicitly, if $\rho_i(g) = 1$ for some i then $\rho_j(g) = 1$ for all j, and hence $g = 1$, and so it follows that the representations ρ_i are faithful. In particular, the subgroup A has a faithful irreducible representation of degree at most n. The fact that $A/Z(G)$ is abelian implies that A is 2-step nilpotent, and so Lemma 9.B.3 implies that $[A : Z(A)] \leq n^2$. Combined with (9.B.3), this implies that

$$[A : Z] \leq n^2, \tag{9.B.4}$$

as required.

We now claim that, more than (9.B.3), we actually have

$$C_G(A) = Z. \tag{9.B.5}$$

If, on the contrary, $C_G(A) \neq Z$ then since $C_G(A)$ is soluble we may let H be the smallest term in the derived series of $C_G(A)$ that is not contained in Z. We then have $H \lhd G$ by Lemmas 1.5.2 and 1.5.3, and HZ/Z abelian. It then follows that HA is a normal subgroup of G, and then since A/Z and HZ/Z are both abelian and $H \subset C_G(A)$ it follows that HA/Z is abelian. By maximality of A we therefore have $H \subset A$; in particular, since $H \subset C_G(A)$ by definition we have $H \subset Z(A)$. However, by (9.B.3) this contradicts the definition of H as not being contained in Z. It must, therefore, indeed be the case that (9.B.5) holds, as claimed.

Finally, we claim that $A = C_{G/Z}(A)$. To see this, note that for every $x \in A$ and $c \in C_{G/Z}(A)$, the commutator $[x, c]$ lies in Z by definition. It therefore follows from Lemma 5.5.3 that the map $A \times C_{G/Z}(A)$ defined by $(x, c) \mapsto [x, c]$ is a homomorphism in each variable. This implies that for each $c \in C_{G/Z}(A)$ the map

$$\begin{aligned} \varphi_c : A/Z &\to Z \\ xZ &\mapsto [x, c] \end{aligned}$$

is a homomorphism, and then that the map

$$\begin{aligned} \psi : C_{G/Z}(A) &\to \mathrm{Hom}(A/Z, Z) \\ c &\mapsto \varphi_c \end{aligned}$$

is a homomorphism. Since G is irreducible, Corollary 9.A.5 implies that Z is isomorphic to a subgroup of $\mathbb{C}^\times$, and so ψ induces a homomorphism $C_{G/Z}(A) \to \mathrm{Hom}(A/Z, \mathbb{C}^\times)$. However, since A/Z is finite, every element of $\mathrm{Hom}(A/Z, \mathbb{C}^\times)$ has its image in the unit circle, and so $\mathrm{Hom}(A/Z, \mathbb{C}^\times)$ is isomorphic to the dual group $\widehat{A/Z}$ and ψ in fact defines a homomorphism $C_{G/Z}(A) \to \widehat{A/Z}$. The kernel of this homomorphism is precisely Z by (9.B.5), and so ψ induces an injective homomorphism $C_{G/Z}(A)/Z \to \widehat{A/Z}$. Since A/Z is finite by (9.B.4), it follows from (4.2.2) that $|C_{G/Z}(A)/Z| \le |A/Z|$. Since $A \subset C_{G/Z}(A)$ by definition of A, it follows that $A = C_{G/Z}(A)$, as claimed. This implies in particular that A is the kernel of the action of G on A/Z, and hence that G/A is isomorphic to a subgroup of $\mathrm{Aut}\,(A/Z)$. By (9.B.4), this implies that $[G : A] \le (n^2)!$, as required, and the proposition is proved. □

Proof of Proposition 9.B.1 By Lemma 9.A.12 (ii) there exists a minimal system of imprimitivity $V_1, \dots, V_r$ for G. By definition each $N_G(V_i)$ acts primitively on V_i, and by Lemma 9.A.12 (i) it also acts irreducibly on V_i. The quotient $N_G(V_i)/C_{N_G(V_i)}(V_i)$ therefore acts primitively, irreducibly and faithfully on V_i, and so its centre has index at most $(n^2)!n^2$ by Proposition 9.B.2.

Define

$$N = \bigcap_{i=1}^{r} N_G(V_i),$$

noting that N is a normal subgroup of G satisfying

$$[G : N] \le r! \le n! \tag{9.B.6}$$

by Lemma 9.A.11. The normality of N and Lemma 1.5.2 imply that the centre $Z(N)$ is an abelian normal subgroup of G. We claim, in addition, that

$$[N : Z(N)] \le ((n^2)!n^2)^n, \tag{9.B.7}$$

which will combine with (9.B.6) to prove the proposition.

To prove (9.B.7), first define A_i to be the pullback to $N_G(V_i)$ of the centre of $N_G(V_i)/C_{N_G(V_i)}(V_i)$, so that

$$[N_G(V_i) : A_i] \le (n^2)!n^2. \tag{9.B.8}$$

Next define

$$A = \bigcap_{i=1}^{r} A_i,$$

noting that

$$[N : A] \leq ((n^2)!n^2)^r \leq ((n^2)!n^2)^n \tag{9.B.9}$$

by (9.B.8). By definition of A_i, we have $[N_G(V_i), A_i] \subset C_{N_G(V_i)}(V_i)$, and hence in particular in $[N_G(V_i), A_i] \subset C_G(V_i)$. It follows that

$$\begin{aligned}[N, A] &\subset \bigcap_{i=1}^{r} C_G(V_i) \\ &= C_G(\mathbb{C}^n) \\ &= \{1\}.\end{aligned}$$

In particular, $A \subset Z(N)$. The claim (9.B.7), and hence the proposition, therefore follow from (9.B.9). □

Proof of Theorem 9.1.4 By Lemma 9.A.3 there exist $\alpha \in GL_n(\mathbb{C})$ and irreducible representations $\rho_i : G \to GL_{n_i}(\mathbb{K})$ for some $n_1, \ldots, n_r$ with $n_1 + \cdots + n_r = n$ such that

$$\alpha^{-1} g \alpha = \begin{pmatrix} \rho_1(g) & & * \\ & \ddots & \\ 0 & & \rho_r(g) \end{pmatrix}$$

for every $g \in G$. Proposition 9.B.1 implies that each group $\rho_i(G)$ contains an abelian normal subgroup A_i of index at most $n!((n^2)!n^2)^n$. The irreducibility of ρ_i and the normality of A_i combine with Lemma 9.A.8 to imply that A_i is completely reducible, and then Corollary 9.A.5 implies that A_i is diagonalisable as a subgroup of $GL_{n_i}(\mathbb{C})$. The normal subgroup H of G defined by

$$H = \bigcap_{i=1}^{r} \rho_i^{-1}(A_i)$$

is therefore upper-triangularisable, and since each subgroup A_i has index $n!((n^2)!n^2)^n$ in $\rho_i(G)$, this upper-triangularisable subgroup has index at most $n!^n((n^2)!n^2)^{n^2}$ in G, and the theorem is proved. □

Exercises

9.1 Show that a finite soluble subgroup H of $GL_n(\mathbb{C})$ has an abelian subgroup of index at most $O_n(1)$. *Hint: First check that the commutator subgroup of* $\mathrm{Upp}_n(\mathbb{C})$ *has no torsion, and then use this to show that a finite subgroup of* $\mathrm{Upp}_n(\mathbb{C})$ *is abelian.*

9.2 Given $a \in \mathbb{C}^\times$ and $b \in \mathbb{C}$, define the map $f_{a,b} : \mathbb{C} \to \mathbb{C}$ via $f_{a,b}(z) = az + b$. Let $G = \{f_{a,b} : a \in \mathbb{C}^\times, b \in \mathbb{C}\}$ be the group of these maps under composition. Write $\pi : G \to G/[G,G]$ for the quotient homomorphism.

(a) Verify that G is indeed a group.

(b) Show that $[G,G] \cong (\mathbb{C},+)$ and $G/[G,G] \cong \mathbb{C}^\times$. For a given $f_{a,b} \in G$, write $\pi(f_{a,b}) \in \mathbb{C}^\times$ under this identification explicitly in terms of a and b. *Hint: First check that $f_{a,b} \circ f_{a',b'} = f_{aa',x}$ for some x depending on a, a', b, b', and that $[f_{a,1}, f_{1,b}] = f_{1,b(1-a^{-1})}$. Then identify $\{[f,g] : f,g \in G\}$, the set of commutators in G.*

(c) Let $K \geq 2$, and suppose that $A \subset G$ is a finite K-approximate group generating a non-abelian subgroup of G. Show that $|\pi(A)| \leq K^{O(1)}$. *Hint: First check that $f_{a,b} \circ f_{1,c} \circ f_{a,b}^{-1} = f_{1,ac}$.*

(d) Deduce that if $A \subset G$ is a finite K-approximate group generating a non-abelian subgroup of G then there is an abelian progression P of size at most $\exp(K^{O(1)})|A|$ and rank at most $K^{O(1)}$ such that A is covered by at most $K^{O(1)}$ left translates of P.

9.3 Let $K \geq 2$. Let $G < GL_n(\mathbb{C})$ be a soluble subgroup, and suppose that $A \subset G$ is a finite K-approximate group.

(a) Show that there is a finite subgroup H and a nilprogression P_{nil} of rank at most $K^{O_n(1)}$ such that $\langle HP_{\mathrm{nil}} \rangle$ is nilpotent of step at most n, such that A is contained in the union of at most $K^{O_n(1)}$ left translates of HP_{nil}, and such that $|HP_{\mathrm{nil}}| \leq \exp(K^{O_n(1)})|A|$.

(b) There is a folklore result stating that if N is a nilpotent subgroup of $\mathrm{Upp}_n(\mathbb{C})$ then there is a torsion-free nilpotent group Γ of the same step as N such that N is isomorphic to a subgroup of $(\mathbb{R}^n/\mathbb{Z}^n) \times \Gamma$. Use this result to show that we may take $H = \{1\}$ in (a). *Hint: Identify a finite-index subgroup of $\langle HP_{\mathrm{nil}} \rangle$ with a subgroup of $(\mathbb{R}^n/\mathbb{Z}^n) \times \Gamma$. Show that A is contained in a union of at most $K^{O_n(1)}$ left translates of an approximate subgroup $B \subset [-\frac{1}{8}, \frac{1}{8}]^n \times \Gamma$, and exhibit a Freiman 3-homomorphism from B to the torsion-free nilpotent group $\mathbb{R}^n \times \Gamma$.*

9.4 Let G be a completely reducible soluble subgroup of $GL_n(\mathbb{C})$. Show that G contains a normal abelian subgroup of index $O_n(1)$. *This generalises Proposition 9.B.1 to completely reducible soluble subgroups of $GL_n(\mathbb{C})$.*

9.5 Show that if $G \subset GL_n(\mathbb{C})$ is irreducible then $Z(G)$ is cyclic.

10

Arbitrary Approximate Subgroups of $GL_n(\mathbb{C})$

10.1 Introduction

The main aim of this chapter is to generalise Theorem 9.1.3 to arbitrary approximate subgroups of $GL_n(\mathbb{C})$, as follows.

Theorem 10.1.1 *Let $K \geq 2$, and suppose that $A \subset GL_n(\mathbb{C})$ is a finite K-approximate group. Then there is a nilpotent subgroup $N < G$ of step at most n such that A is contained in a union of at most* $\exp(K^{O_n(\log K)})$ *left cosets of N.*

Remark 10.1.2 The assumption $K \geq 2$ is for notational convenience, as described in Remark 6.1.2. For $K < 2$, a K-approximate group is an exact group, and in that case Theorem 10.1.1 reduces to a well-known theorem of Jordan stating that a finite subgroup of $GL_n(\mathbb{C})$ contains an abelian subgroup of index at most $O_n(1)$. Theorem 10.1.1 can thus be seen as an approximate-group analogue of Jordan's theorem.

We reduce Theorem 10.1.1 to Theorem 9.1.3 by showing that A is contained in the union of at most $\exp(K^{O_n(\log K)})$ left translates of a K^3-approximate group $B \subset A^2$ that generates a soluble subgroup of $GL_n(\mathbb{C})$. This reduction was first proved by Hrushovski [42] with ineffective bounds, and then proved with partially effective bounds by Breuillard, Green and Tao [16] and then with effective bounds by Pyber and Szabó [49]. Here we present yet another proof, also due to Breuillard, Green and Tao [17].

Unlike almost all of the other results we prove in this book, we do not offer a self-contained proof of Theorem 10.1.1. In particular, in reducing Theorem 10.1.1 to Theorem 9.1.3 we rely on a substantial result of Breuillard called the *uniform Tits alternative*, which we state below as Theorem 10.2.2. We also assume a result of Mal'cev and Platonov, which we state below as Proposition 10.2.1. Assuming these results allows us to prove Theorem 10.1.1 without developing any algebraic group theory, which would be far beyond the scope of this book.

One reason for including Theorem 10.1.1 is that it gives the opportunity to introduce two tools that have uses beyond the proof of the theorem. The first of these is the Tits alternative itself, which is an important structural result for linear groups that is also relevant to the topic of *growth* that we introduce in Chapter 11. The second is a result due to Sanders and Croot–Sisask, which also features, for example, in the proof of Theorem 7.1.1.

10.2 Free Groups and the Uniform Tits Alternative

Our approach to Theorem 10.1.1 is centred on a well-known phenomenon of linear groups called the *Tits alternative*. First proved by Tits [71], this states that either a finitely generated linear group has a soluble subgroup of finite index or it contains a non-abelian free subgroup. If a group G has a soluble subgroup of finite index then we call G *virtually soluble*. Indeed, more generally, given an adjective $\mathcal{P}$ that might be applied to a group, we say that a group is *virtually* $\mathcal{P}$ if it has a finite-index subgroup that is $\mathcal{P}$.

The rough idea of the reduction of Theorem 10.1.1 to Theorem 9.1.3 is to show that a large piece of A does not contain any generating set of a free group, and then apply the Tits alternative to show that this piece must generate a virtually soluble subgroup and give us some hope of applying Theorem 9.1.3.

Applying the Tits alternative as stated above presents some problems. First, we need control over the index of the soluble subgroup in order for it to be useful. This control is provided by the following result of Mal'cev and Platonov.

Proposition 10.2.1 (Mal'cev–Platonov [75, Corollary 10.11]) *Let G be a virtually soluble subgroup of $GL_n(\mathbb{C})$. Then G contains a soluble subgroup H of index $O_n(1)$.*

Remark Combined with Exercise 9.1, Proposition 10.2.1 recovers Jordan's theorem for finite linear groups (see Remark 10.1.2). We caution, however, that Jordan's theorem is an ingredient in the proof of Proposition 10.2.1.

Second, even if A itself does not contain a generating set of a free subgroup, in principle it could be that some large power A^m of A contains a generating set for a free subgroup, in which case A will not be virtually soluble. We eliminate this possibility with the following strengthening of the Tits alternative, due to Breuillard.

Theorem 10.2.2 (uniform Tits alternative; Breuillard [11]) *Let $A \subset GL_n(\mathbb{C})$ be a finite symmetric set containing the identity. Then either $\langle A \rangle$ is virtually soluble or $A^{O_n(1)}$ contains two generators of a non-abelian free subgroup of $GL_n(\mathbb{C})$.*

10.3 Small Neighbourhoods of the Identity

As we invite the reader to show in Exercise 10.1, one consequence of Theorem 4.1.3 is that if A is a finite symmetric subset of an abelian group G satisfying $|A + A| \leq K|A|$, then for each $m \in \mathbb{N}$ there exists a finite symmetric set $S \subset G$ containing the identity such that $|S| \geq m^{-O(K^{O(1)})}|A|$ and $mS \subset 4A$. The following result of Sanders [58] and Croot and Sisask [23] extends this to sets of small doubling in arbitrary groups, via an elementary proof that does not use any heavy machinery along the lines of Theorem 4.1.3.

Proposition 10.3.1 (Sanders; Croot–Sisask) *Let A be a finite symmetric set containing the identity in a group G, and suppose that $|A^2| \leq K|A|$. Let $m \in \mathbb{N}$. Then there exists a symmetric set S containing the identity with $|S| \geq \exp(-K^{O(m)})|A|$ such that $S^m \subset A^4$.*

Proposition 10.3.1 is an ingredient in our proof of Theorem 10.1.1, but its use goes beyond that as it is also an important ingredient in the proof of Theorem 7.1.1.

Before we prove Proposition 10.3.1 we record the following simple lemma.

Lemma 10.3.2 *Let X be a set and let $n \in \mathbb{N}$. Then we may define a metric d on the subspaces of X of size n via $d(A, B) = |A \triangle B|$.*

Proof This is double the metric for the graph on the subspaces of X of size n in which A and B are adjacent if and only if there exist $a \in A$ and $x \in X$ such that $B = (A \setminus \{a\}) \cup \{x\}$. □

Proof of Proposition 10.3.1 We follow [18, Theorem 5.3]. Define a function

$$f : (0,1] \to \mathbb{R}$$
$$\lambda \mapsto \min\left\{\frac{|AB|}{|A|} : B \subset A, |B| \geq \lambda|A|\right\},$$

noting that $1 \leq f(\lambda) \leq K$ for every $\lambda \in (0,1]$ because $|A^2| \leq K|A|$. We claim that there exists $\lambda \geq \exp(-K^{O(m)})$ such that

$$f\left(\frac{\lambda^2}{2K}\right) \geq \left(1 - \frac{1}{2m}\right) f(\lambda). \tag{10.3.1}$$

Indeed, letting n be minimal such that (10.3.1) is satisfied for $\lambda = \left(\frac{1}{2K}\right)^{2^{n+1}-1}$ we see that

$$f\left(\left(\frac{1}{2K}\right)^{2^{r+1}-1}\right) \geq \left(1 - \frac{1}{2m}\right) f\left(\left(\frac{1}{2K}\right)^{2^{r}-1}\right)$$

for every $r \leq n$. Since $f(1) \leq K$, this in turn implies that

$$f\left(\left(\frac{1}{2K}\right)^{2^{n+1}-1}\right) < \left(1 - \frac{1}{2m}\right)^{n} K,$$

and since $f(\lambda) \geq 1$ for every λ this forces

$$n \leq -\frac{\log K}{\log(1 - \frac{1}{2m})} \ll m \log K,$$

so that (10.3.1) is satisfied in particular for

$$\begin{aligned}\lambda = \left(\frac{1}{2K}\right)^{2^{n+1}-1} &\geq K^{-2^{O(m \log K)}} \\ &\geq K^{-K^{O(m \log 2)}} \geq \exp(-K^{O(m)}),\end{aligned}$$

as claimed.

Now fix a set B attaining the minimum in the definition of $f(\lambda)$; thus

$$|B| \geq \lambda|A| \tag{10.3.2}$$

and

$$|AB| = f(\lambda)|A|. \tag{10.3.3}$$

Since $|Ba| = |B|$ for every $a \in A$ we have $\sum_{x \in G} \sum_{a \in A} 1_{Ba}(x) = |A||B|$.

However, since $Ba \subset A^2$ this sum is only non-zero when $x \in A^2$, so we in fact have

$$\sum_{x\in A^2}\sum_{a\in A} 1_{Ba}(x) = |A||B|,$$

which by the Cauchy–Schwarz inequality (1.5.2) implies that

$$\sum_{x\in A^2}\left(\sum_{a\in A} 1_{Ba}(x)\right)^2 \geq \frac{|A|^2|B|^2}{|A^2|}. \tag{10.3.4}$$

Now the left-hand side of (10.3.4) is equal to

$$\sum_{x\in A^2}\sum_{a,a'\in A} 1_{Ba\cap Ba'}(x) = \sum_{a,a'\in A}|Ba\cap Ba'|,$$

and so (10.3.4) and the pigeonhole principle imply that there exists $a_0 \in A$ such that

$$\sum_{a\in A}|Ba\cap Ba_0| \geq \frac{|A||B|^2}{|A^2|} \geq \frac{1}{K}|B|^2 \geq \frac{\lambda^2|A|^2}{K}, \tag{10.3.5}$$

the last inequality following from (10.3.2).

Write

$$C = \left\{a \in A : |Ba\cap Ba_0| \geq \frac{\lambda^2|A|}{2K}\right\},$$

and note that

$$\begin{aligned}\frac{\lambda^2|A|^2}{K} &\leq \sum_{a\in C}|Ba\cap Ba_0| + \sum_{a\in A\backslash C}|Ba\cap Ba_0| \qquad \text{(by (10.3.5))}\\ &\leq |C||B| + |A|\frac{\lambda^2|A|}{2K}\\ &\leq |A|\left(|C| + \frac{\lambda^2|A|}{2K}\right),\end{aligned}$$

and hence that

$$|C| \geq \frac{\lambda^2|A|}{2K} \geq \exp(-K^{O(m)})|A|. \tag{10.3.6}$$

Set $S = a_0^{-1}C \cup C^{-1}a_0 \cup \{1\}$, noting that S is symmetric and contains the identity. We claim that for every $s \in S$ we have

$$|Bs\cap B| \geq \frac{\lambda^2|A|}{2K}. \tag{10.3.7}$$

If $s = 1$ this is immediate from (10.3.2). If $s = a^{-1}a_0$ for some $a \in C$

then $(Ba \cap Ba_0)a^{-1} \subset B \cap Ba_0a^{-1} = Bs \cap B$, and so $|Bs \cap B| \geq Ba \cap Ba_0$ and the claim follows from the definition of C. Similarly, if $s = a_0^{-1}a$ for some $a \in C$ then $(Ba \cap Ba_0)a_0^{-1} \subset Ba_0^{-1}a \cap B = Bs \cap B$ and again the claim follows from the definition of C.

The definition of f combines with (10.3.1) and (10.3.7) to imply that for every $s \in S$ we have $|A(Bs \cap B)| \geq (1 - \frac{1}{2m})f(\lambda)|A|$, and hence

$$\begin{aligned}|ABs \cap AB| &\geq |A(Bh \cap B)| \\ &\geq \left(1 - \frac{1}{2m}\right) f(\lambda)|A| \geq \left(1 - \frac{1}{2m}\right)|AB|\end{aligned}$$

by (10.3.3). This means in particular that $|ABs \triangle AB| \leq \frac{1}{m}|AB|$ for every $s \in S$, and hence, by Lemma 10.3.2 and the triangle inequality, that $|ABx \triangle AB| \leq |AB|$ for every $x \in S^m$, which means that

$$|ABx \cap AB| \geq \frac{|AB|}{2}$$

for every $x \in S^m$. However, this means in particular that $|ABx \cap AB| \geq 1$ for every $x \in S^m$, and hence that $S^m \subset A^4$. We also have $|S| \geq \exp(-K^{O(m)})|A|$ by (10.3.6), and so the proposition is satisfied. □

10.4 Approximate Subgroups of Complex Linear Groups

In this section we complete the proof of Theorem 10.1.1. Showing that a large piece of A does not contain a generating set of a free group does not turn out to be too difficult. The spirit of this statement is captured by the following result, which shows that generating sets of free groups always force a certain amount of growth.

Lemma 10.4.1 *Let G be a group, and suppose that $x_1, x_2 \in G$ generate a non-abelian free subgroup. Write $X = \{x_1^{\pm 1}, x_2^{\pm 1}\}$. Let $A \subset G$ be finite. Then $|AX| \geq 3|A|$.*

Proof Let T be a left transversal for the free group $F = \langle x_1, x_2 \rangle$ in G with $1 \in T$, so that $G = TF$. For each $g \in G$ let $\tau(g) \in T$ and $\varphi(g) \in F$ be the unique elements such that $g = \tau(g)\varphi(g)$. Note that $\varphi(gy) = \varphi(g)y$ for every $g \in G$ and $y \in F$. This implies that, given $g \in G$, for three of the four choices of $y \in X$ the reduced-word representation of $\varphi(gy)$ is precisely the reduced-word representation of $\varphi(g)$ with an additional

y on the right. Given $g \in G$, write $U(g)$ for the set of elements gy as y ranges over these three choices of $y \in X$.

We claim that $U(g) \cap U(h) = \varnothing$ whenever g and h are distinct elements of G. Indeed, given $u \in U(g)$, define $w(u)$ to be the element of F obtained by deleting the right-most element from the reduced-word representation of $\varphi(u)$, noting that $w(u)$ is well defined by definition of $U(g)$. We then have $g = \tau(u)w(u)$, meaning we can recover g from u, which proves the claim. Since $AX \supset \bigcup_{a \in A} U(a)$, the claim in turn implies the lemma. □

It is the next result that allows us to apply Lemma 10.4.1 to K-approximate groups with $K > 3$.

Lemma 10.4.2 ([17, Proposition 2.2]) *Let A be a finite K-approximate group, and let $m \in \mathbb{N}$. Then there exist symmetric subsets $B \subset A^5$ and $S \subset A^4$ containing the identity with $|B| \geq |A|$ and*

$$|S| \geq \exp(-K^{O(m \log K)})|A|$$

such that $|BS^m| \leq 2|B|$.

Proof Set $r = \lceil 4 \log_2 K \rceil$. It follows from Proposition 10.3.1 that there exists a symmetric set S containing the identity with

$$|S| \geq \exp(-K^{O(m \log K)})|A|$$

such that $S^m \subset A^4$. Since

$$A \subset AS \subset AS^2 \subset \cdots \subset AS^{rm} \subset A^5$$

and $|A^5| \leq K^4|A| \leq 2^r|A|$, there must exist $j \in \mathbb{Z}$ with $0 \leq j < r$ such that

$$|AS^{(j+1)m}| \leq 2^r|AS^{jm}|.$$

The lemma then follows with $B = AS^{jm}$. □

Proof of Theorem 10.1.1 We follow Breuillard, Green and Tao [17, Theorem 1.3]. Applying Lemma 10.4.2 with $m \in \mathbb{N}$ the constant implied by the notation $O_n(1)$ in the conclusion of Theorem 10.2.2, we obtain sets $B \subset A^5$ and $S \subset A^4$ containing the identity with $|B| \geq |A|$ and $|S| \geq \exp(-K^{O_n(\log K)})|A|$ such that $|BS^m| \leq 2|B|$. Lemma 10.4.1 therefore implies that S^m does not contain any pair of generators for a non-abelian free subgroup of $GL_n(\mathbb{C})$, and so Theorem 10.2.2 implies that $\langle S \rangle$ is virtually soluble. Proposition 10.2.1 therefore implies that $\langle S \rangle$

has a soluble subgroup H of index $O_n(1)$, and so Lemma 2.6.2 implies that

$$\begin{aligned}|A^8 \cap H| &\gg_n |A^4 \cap S| \\ &= |S| \\ &\geq \exp(-K^{O_n(\log K)})|A|.\end{aligned}$$

Lemma 7.1.5 therefore implies that there exists $X \subset GL_n(\mathbb{C})$ of size at most $\exp(K^{O_n(\log K)})$ such that $A \subset X(A^2 \cap H)$. The set $A^2 \cap H$ is a K^3-approximate group by Proposition 2.6.5, and so the desired result follows from Theorem 9.1.3. □

Exercises

10.1 Use Theorem 4.1.3 to show that if A is a finite symmetric subset of an abelian group G satisfying $|A + A| \leq K|A|$ then for each $m \in \mathbb{N}$ there exists a finite symmetric set $S \subset G$ containing the identity such that $|S| \geq m^{-O(K^{O(1)})}|A|$ and $mS \subset 4A$. *This improves Proposition 10.3.1 for abelian groups, albeit by appealing to a substantial theorem.*

11

Applications to Growth in Groups

11.1 Introduction

In this chapter we introduce some applications of approximate groups to the topic of *growth* in groups. Growth is an important area of study, with links to random walks in groups, differential geometry and geometric group theory, and approximate groups turn out to provide a powerful new means to investigate it.

Let G be a finitely generated group, and let S be a finite symmetric generating set containing the identity. The *growth* of G with respect to S refers to the rate at which the cardinalities $|S^n|$ grow as $n \to \infty$. When $G = \mathbb{Z}^d$ with S the standard generating set, it is easy to see that there exist constants c and C depending on d such that

$$cn^d \leq |nS| \leq Cn^d$$

(since we write $\mathbb{Z}^d$ additively we have nS in place of S^n here). On the other hand, if S is the standard generating set for the free group F_2 then the reader is invited to show in Exercise 11.2 that $|S^n| = 2 \cdot 3^n - 1$ for every $n \in \mathbb{N}$. These different possible growth rates are captured in the following definitions.

Definition 11.1.1 (polynomial, exponential and intermediate growth) Let G be a finitely generated group, and let S be a finite symmetric generating set containing the identity.

- We say that G has *polynomial growth* (with respect to S) if there exist C, d such that $|S^n| \leq Cn^d$ for all $n \in \mathbb{N}$; in this case we refer to d as the *degree* of the polynomial growth.
- We say that G has *exponential growth* (with respect to S) if there exists $\alpha > 1$ such that $|S^n| \geq \alpha^n$ for all $n \in \mathbb{N}$.

- We say that G has *intermediate growth* (with respect to S) if it has neither polynomial nor exponential growth.

Remarks

(i) The assumptions that S is symmetric and contains the identity, which we make throughout this chapter, are not essential for the theory but are notationally convenient. A particularly useful consequence is that

$$S^1 \subset S^2 \subset \dots$$

and $S^n \nearrow G$. When G is finite, a useful further consequence is that there is some n such that $S^n = G$; this need not be the case if S does not contain the identity, for example if $G = \mathbb{Z}/2\mathbb{Z}$ and $S = \{1\}$.

(ii) Note that $|S^n| \leq |S|^n$ for every subset S of a group and every $n \in \mathbb{N}$, so that there is always an exponential upper bound on the growth of any group.

(iii) We have seen straightforward examples of groups of polynomial and exponential growth, but it is not obvious a priori that there should exist groups of intermediate growth. It turns out that these do exist, the first examples being the famous *Grigorchuk groups*. We do not discuss them further in this book, but the interested reader may consult [25, Chapter 8] for details.

The first thing to note about the notions given in Definition 11.1.1 is that they do not depend on the choice of generating set, as follows.

Lemma 11.1.2 *If a group G has polynomial growth of degree d with respect to some finite generating set then it has polynomial growth of degree d with respect to all finite generating sets. If it has exponential growth with respect to some finite generating set then it has exponential growth with respect to all finite generating sets.*

Proof Let S_1 and S_2 be two generating sets for G. Since S_1 is a generating set there exists $k \in \mathbb{N}$ such that $S_2 \subset S_1^k$, and hence $S_2^n \subset S_1^{kn}$ for every $n \in \mathbb{N}$. In particular, if $|S_1^n| \leq Cn^d$ then $|S_2^n| \leq (Ck^d)n^d$, and so polynomial growth of degree d with respect to S_1 implies it with respect to S_2. On the other hand, if $|S_2^n| \geq \alpha^n$ for every $n \in \mathbb{N}$ then $|S_1^{kn}| \geq \alpha^n$ for every $n \in \mathbb{N}$, which implies in particular that $|S_1^n| \geq (\alpha^{1/2k})^n$ for every $n \geq k$ and $|S_1^n| \geq 2 \geq (2^{1/k})^n$ for every $n \leq k$. Thus exponential growth with respect to S_2 implies it with respect to S_1. □

We may thus drop the phrase 'with respect to S' when referring to the growth of a group.

In the next section we will also show that the notions of Definition 11.1.1 are stable under passing between a group and a subgroup of finite index, as follows.

Proposition 11.1.3 *Let G be a group and H a finite-index subgroup. Then H is finitely generated if and only if G is. Moreover, in the event that G and H are finitely generated, H has polynomial growth of degree d if and only if G does, and exponential growth if and only if G does.*

In this book we will be mainly concerned with groups of polynomial growth, which arise naturally in various contexts, such as random walks and non-negatively curved manifolds. It is straightforward to show that nilpotent groups have polynomial growth, as follows.

Proposition 11.1.4 *Let G be an s-step nilpotent group, and suppose that $X \subset G$ has size at most $r \in \mathbb{N}$. Then $|X^n| \leq n^{O_{r,s}(1)}$ for every $n \geq 2$.*

Proof We may assume that $|X| = r$ and write $X = \{x_1, \ldots, x_r\}$. Write $c_1, \ldots, c_d$ for the ordered list of basic commutators in the x_i of total weight at most s. An element $y \in X^n$ can be expressed as a string of length n in the elements $x_1, \ldots, x_r$, but not their inverses. We apply the collecting process of Section 5.4 to such a string in order to write y in the form $c_1^{\ell_1} \cdots c_d^{\ell_d}$ with $\ell_i \in \mathbb{Z}$.

This process uses only transformations of type 1, and results in only positive powers of basic commutators, so $\ell_i \geq 0$ for each i. Moreover, a basic commutator $c_k = [c_j, c_i]$ arising from this process can result only from interchanging an instance of c_j with an instance of c_i, and each such pair will be interchanged at most once. It therefore follows by induction on k that $\ell_k \leq n^{|\chi(c_k)|}$ for each k. In particular, we have

$$X^n \subset \{c_1^{\ell_1} \cdots c_d^{\ell_d} : 0 \leq \ell_i \leq n^{|\chi(c_i)|}\},$$

and the proposition follows. □

It then follows from Proposition 11.1.3 that any group with a finite-index nilpotent subgroup has polynomial growth (recall from Section 10.2 that we call such groups *virtually nilpotent*). Remarkably, it turns out that the converse is also true, thanks to the following theorem of Gromov.

Theorem 11.1.5 (Gromov [36]) *For every $C > 0$ and $d \geq 0$ there exists $N = N_{C,d} \in \mathbb{N}$ such that if G is a group with a finite symmetric generating set S containing the identity and satisfying $|S^n| \leq Cn^d$ for each $n = 1, \ldots, N$ then G contains an $O_{C,d}(1)$-step nilpotent subgroup of index at most $O_{C,d}(1)$. In particular, if G is a finitely generated group of polynomial growth then G is virtually nilpotent.*

Approximate groups arise quite naturally in the study of polynomial growth; indeed, as we shall see in Lemma 11.3.2, if a group has polynomial growth of degree d with respect to the generating set S then there exist infinitely many $n \in \mathbb{N}$ for which S^n is an $O_d(1)$-approximate group. Building on work of Hrushovski [42], Breuillard, Green and Tao [18] exploited this and Theorem 7.1.1 to obtain the following version of Gromov's theorem.

Theorem 11.1.6 (Breuillard–Green–Tao [18, Corollary 11.5]) *Given $d \geq 0$ there exists $N = N_d$ such that if G is a group generated by a finite symmetric set S containing the identity, and if*

$$|S^n| \leq n^d |S| \tag{11.1.1}$$

for some $n \geq N$, then G is virtually $O_d(1)$-step nilpotent.

The main aim of this chapter is to prove Theorem 11.1.6, which we do in Section 11.3. By far the most substantial ingredient is Theorem 7.1.1, which we do not prove in this book, and so the proof of Theorem 11.1.6 we give is a long way from self-contained. We do, though, give complete details of its deduction from Theorem 7.1.1.

In Sections 11.4 and 11.6 we present some applications of Theorem 11.1.6.

Remarks There are two key points differentiating Theorem 11.1.6 from Theorem 11.1.5. The first point is that in Theorem 11.1.6 the hypothesis of a polynomial bound on $|S^n|$ is only required to hold for a single value of n. The second point is that in Theorem 11.1.6 the generating set S may be arbitrarily large without violating the hypothesis of the theorem. This is in contrast to Theorem 11.1.5, which requires in particular that $|S| \leq C$. The price of this is that whilst Theorem 11.1.5 gives a bound of $O_{C,d}(1)$ on the the index of the nilpotent subgroup, in Theorem 11.1.6 this index can be arbitrarily large. To see that this is unavoidable, consider the case in which H is a large non-abelian finite simple group, N is a nilpotent group, $G = H \times N$ and S is the union of H and a generating set for N.

We invite the reader to show in Exercise 11.4 that if one replaces the condition (11.1.1) in Theorem 11.1.6 with $|S^n| \leq Cn^d$ then one can once again bound the index of the nilpotent subgroup in terms of C and d whilst still requiring only that such a bound hold for a single value of n.

11.2 Finite-Index Subgroups

In this section we present a number of results that are useful for comparing a group to its finite-index subgroups. Our main aim is to prove Proposition 11.1.3, but we start with the following simple observation.

Lemma 11.2.1 *Let G be a group with finite symmetric generating set S containing the identity, and let H be a subgroup of index at least $m \in \mathbb{N}$ in G. Then S^{m-1} has non-empty intersection with at least m distinct left cosets of H.*

Proof If $S^{n+1}H = S^nH$ for a given $n \geq 0$ then it follows by induction that $S^rH = S^nH$ for every $r \geq n$, and hence that $G = S^nH$. We may therefore assume that $H \subsetneq SH \subsetneq S^2H \subsetneq \cdots \subsetneq S^{m-1}H$. This implies that the number of left cosets of H having non-empty intersection with S^n is strictly increasing for $n = 0, 1, \dots, m-1$, and so the lemma is proved. □

We now move on to Proposition 11.1.3, which shows that the notions of finite generation and growth are stable under moving between groups and their finite-index subgroups. Throughout we implicitly use Lemma 11.1.2, which says that the growth rate does not depend on the choice of generating set.

Proposition 11.2.2 *Let G be a group, and let H be a subgroup of index $k \in \mathbb{N}$. Suppose that S is a finite symmetric generating set for G containing the identity, and let $X \subset S^k$ be the complete set of left-coset representatives for H in G given by Lemma 11.2.1. Then for every $n \geq 2$ we have $S^{nk} \subset X(H \cap S^{3k})^{n-1}$. In particular, $H \cap S^{3k}$ generates H.*

Proof Since X is a complete set of left-coset representatives for N we have $S^{2k} \subset XH$, and since $X \subset S^k$ this in fact implies that $S^{2k} \subset X(H \cap S^{3k})$, which is the $n = 2$ case of the proposition. For $n > 2$ we then have

$$\begin{aligned} S^{nk} &= S^k S^{(n-1)k} \\ &\subset S^k X (H \cap S^{3k})^{n-2} && \text{(by induction)} \\ &\subset S^{2k} (H \cap S^{3k})^{n-2} && \text{(since } X \subset S^k\text{)} \\ &\subset X (H \cap S^{3k})^{n-1} && \text{(by the case } n = 2\text{),} \end{aligned}$$

as required. □

Proof of Proposition 11.1.3 If H is finitely generated by T, say, and X is a complete set of left-coset representatives for H in G, then G is finitely generated by $T \cup X \cup X^{-1}$ and

$$|T^n| \leq |(T \cup X \cup X^{-1})^n|$$

for every $n \in \mathbb{N}$. Conversely, if G is finitely generated by S, say, and k is the index of H in G, then Proposition 11.2.2 implies that H is finitely generated by $H \cap S^{3k}$ and

$$|S^n| \leq |S^{kn}| \leq k|(H \cap S^{3k})^n|$$

for every $n \in \mathbb{N}$. □

We close this section by noting the following useful trick, which allows us, for example, to assume that the nilpotent subgroup of finite index given by Gromov's theorem is normal.

Lemma 11.2.3 *Let G be a group and let $H < G$ have index $k \in \mathbb{N}$ in G. Then there exists a subgroup $N < H$ with $N \lhd G$ such that $[G : N] \leq k^k$.*

Proof Take $N = \bigcap_{gH \in G/H} gHg^{-1}$; it is easy to check that this is well defined, normal and of the required index in G. □

11.3 A Refinement of Gromov's Theorem

The relevance of approximate groups to Gromov's theorem comes from the following fact.

Proposition 11.3.1 *Let $d > 0$. Suppose that G is a group and $S \subset G$ is a finite symmetric subset containing the identity and satisfying $|S^n| \leq n^d|S|$ for some $n \geq 3^{12}$. Then there exists $m \in \mathbb{N}$ with $n^{1/2} \leq m \leq 2n^{5/6}$ such that S^m is an $e^{O(d)}$-approximate group.*

The main content of Proposition 11.3.1 is the following lemma.

Lemma 11.3.2 *Let $d > 0$ and let $q \in \mathbb{N}$. Then there exists $K = K_{d,q}$ such that if G is a group and $S \subset G$ is a finite symmetric subset containing the identity and satisfying $|S^n| \leq n^d|S|$ for some $n \geq q^{12}$, then there exists $m \in \mathbb{N}$ with $\lfloor n^{1/2} \rfloor \leq m \leq n^{5/6}$ such that $|S^{qm}| \leq K|S^m|$. Indeed, we may take $K = q^{4d}$.*

Proof The assumption that $n \geq q^{12}$ implies that $n^{1/12} \geq q$, and in particular that there exists some $r \in \mathbb{N}$ such that

$$n^{3/4} \leq q^r \lfloor n^{1/2} \rfloor \leq n^{5/6}. \tag{11.3.1}$$

Suppose the lemma does not hold for a given value of K. This means in particular that $|S^{q^k \lfloor n^{1/2} \rfloor}| > K|S^{q^{k-1} \lfloor n^{1/2} \rfloor}|$ for every $k = 1, \ldots, r$, and hence that $|S^{\lfloor n^{5/6} \rfloor}| > K^r |S^{\lfloor n^{1/2} \rfloor}|$. However, it follows from (11.3.1) that $r \geq \frac{1}{4} \log_q n$ (that is to say, $r \geq \log_q n^{1/4}$), and hence that

$$\begin{aligned} |S^n| &\geq |S^{\lfloor n^{5/6} \rfloor}| \\ &> K^{\frac{1}{4}\log_q n} |S^{\lfloor n^{1/2} \rfloor}| \\ &= n^{\frac{1}{4}\log_q K} |S^{\lfloor n^{1/2} \rfloor}| \\ &\geq n^{\frac{1}{4}\log_q K} |S|, \end{aligned}$$

which by the hypothesis of the lemma implies that $K < q^{4d}$. The lemma therefore holds with $K = q^{4d}$, as claimed. □

Proof of Proposition 11.3.1 Applying Lemma 11.3.2 with $q = 3$ gives $m \in \mathbb{N}$ with $\lfloor n^{1/2} \rfloor \leq m \leq n^{5/6}$ such that $|S^{3m}| \leq 3^{4d}|S^m| = e^{O(d)}|S^m|$. Proposition 2.5.5 then implies that S^{2m} is an $e^{O(d)}$-approximate group, as required. □

Combining Proposition 11.3.1 with Theorem 7.1.1, it is fairly easy to arrive at the following statement in the direction of Theorem 11.1.6.

Proposition 11.3.3 *Given $d > 0$, there exists $N = N_d$ such that if G is a group generated by a finite symmetric set S containing the identity, and if*

$$|S^n| \leq n^d|S| \tag{11.3.2}$$

for some $n \geq N$, then there exist a subgroup $C < G$ of index at most $O_d(1)$ in G and a subgroup $H \lhd C$ with $H \subset S^{\lfloor n/2 \rfloor}$ such that C/H is $O_d(1)$-step nilpotent. Moreover, there exists $n_0 \in \mathbb{N}$ with $n^{1/2} \leq n_0 \leq \frac{1}{2}n$ such that S^{n_0} is an $e^{O(d)}$-approximate group.

Proof Let N be a constant to be determined shortly, and suppose that (11.3.2) holds for some $n \geq N$. Provided $N > 3^{12}$, Proposition 11.3.1 implies that there exists $n_0 \in \mathbb{N}$ with

$$n^{1/2} \leq n_0 \leq 2n^{5/6} \tag{11.3.3}$$

such that S^{n_0} is an $e^{O(d)}$-approximate group. Theorem 7.1.1 then gives a constant $R = R_d \in \mathbb{N}$ and subgroups $H \lhd C < G$ with

$$H \subset S^{4n_0} \tag{11.3.4}$$

such that C/H is $O_d(1)$-step nilpotent and such that S^{n_0} is contained in the union of at most R left cosets of C. Provided $N \geq R^2$, (11.3.3) then implies that S^R is contained in the union of at most R left cosets of C, which by Lemma 11.2.1 implies that the index of C in G is at most R. Finally, provided $N \geq 16^6$, we have $2n^{5/6} \leq \frac{1}{8}n$, and so (11.3.3) implies that $n_0 \leq \frac{1}{8}n$ and (11.3.4) implies that $H \subset S^{\lfloor n/2 \rfloor}$, as required. □

The following lemma shows how to pass from Proposition 11.3.3 to Theorem 11.1.6.

Lemma 11.3.4 *Let G be a group with a finite normal subgroup H, and suppose that G/H is s-step nilpotent. Then G has a subgroup of index at most $|H|!$ that is $(s+1)$-step nilpotent.*

Proof The group G acts by conjugation on H, and the orbit-stabiliser theorem applied to this action implies that $C_G(H)$ has index at most $|H|!$ in G. Since $C_G(H)/(C_G(H) \cap H)$ is nilpotent of step s and $(C_G(H) \cap H)$ is central in $C_G(H)$, the subgroup $C_G(H)$ is $(s+1)$-step nilpotent, and the lemma is proved. □

Proof of Theorem 11.1.6 Apply Proposition 11.3.3. Since H is finite, Lemma 11.3.4 implies that C is virtually $O_d(1)$-step nilpotent, and then since $[G : C] \leq \infty$ it follows that G itself is virtually $O_d(1)$-step nilpotent, as required. □

11.4 Persistence of Polynomial Growth

Given a group G with a finite symmetric generating set S containing the identity, if $|S^n| \leq n^d|S|$ for some $n \in \mathbb{N}$ and $d \geq 0$ we say that G has *polynomial growth of degree d at scale n* with respect to S. Theorem 11.1.6 shows that polynomial growth of degree d of a group G

at a single large enough scale n implies that G is virtually nilpotent. We know from Propositions 11.1.4 and 11.1.3 that this in turn implies that G exhibits polynomial growth of some degree d' at all scales $m \geq n$. The following corollary of Theorem 11.1.6 shows that we can bound the degree d' of this subsequent polynomial growth in terms of d.

Corollary 11.4.1 (polynomial growth at one sufficiently large scale implies polynomial growth at all subsequent scales; Breuillard–Green–Tao [18, Corollary 11.9]) *Given $d > 0$ there exists $N = N_d$ such that if G is a group generated by a finite symmetric set S containing the identity and if*

$$|S^n| \leq n^d|S| \tag{11.4.1}$$

for some $n \geq N$ then $|S^m| \leq m^{O_d(1)}|S|$ for every $m \geq n$.

Remark 11.4.2 Tao [67, Example 1.11] gives an example to show that polynomial growth at subsequent scales given by Corollary 11.4.1 can be of higher degree than d. Indeed, if

$$S = \begin{pmatrix} 1 & [-n,n] & [-n^3,n^3] \\ 0 & 1 & [-n,n] \\ 0 & 0 & 1 \end{pmatrix} \subset \begin{pmatrix} 1 & \mathbb{Z} & \mathbb{Z} \\ 0 & 1 & \mathbb{Z} \\ 0 & 0 & 1 \end{pmatrix}$$

then S satisfies $|S^n| \leq n^3|S|$ regardless of the choice of n, but for any fixed n we have $|S^m| \gg m^4$ as $m \to \infty$ (note that, although S is not symmetric, this can be fixed by considering the set $S \cup S^{-1}$ in its place; we leave the details to the reader). Tessera and the author [69, Theorem 1.11] have nonetheless shown that the degree of polynomial growth at subsequent scales can be bounded by d if the assumption $|S^n| \leq n^d|S|$ is replaced with the stronger hypothesis $|S^n| \leq Cn^d$, confirming a conjecture of Benjamini.

The main difficulty of applying Theorem 11.1.6 in the proof of Corollary 11.4.1 is to obtain a bound on $|S^m|$ that is uniform over all groups G and generating sets S satisfying (11.4.1). To overcome this we use the following extension of Proposition 11.1.3

Proposition 11.4.3 *Let G be a group with an s-step nilpotent subgroup N of index at most $k \in \mathbb{N}$, and suppose that X is a subset of G with $|X| \leq r$. Then $|X^n| \leq n^{O_{k,r,s}(1)}$ for all $n \geq 2$.*

Proof Since $[\langle X\rangle : \langle X\rangle \cap N] \leq [G : N]$, we may assume for notational convenience that $\langle X\rangle = G$. Set $S = X \cup X^{-1} \cup \{1\}$. Lemma 11.2.1 implies that S^k contains a complete set Z of left-coset representatives for N in G. Proposition 11.2.2 implies that for every $n \in \mathbb{N}$ we have

$S^{nk} \subset Z(N \cap S^{3k})^{n-1}$, and hence $|S^{nk}| \le k|(N \cap S^{3k})^{n-1}|$. Since $|N \cap S^{3k}| \le (3r)^{3k}$, we therefore have $|X^n| \le |S^n| \le |S^{nk}| \le kn^{O_{r,k,s}(1)}$ by Proposition 11.1.4, and the result follows. □

Proof of Corollary 11.4.1 Proposition 11.3.3 and Lemma 11.2.3 give n_0 with

$$n^{1/2} \le n_0 \le \tfrac{1}{2}n \tag{11.4.2}$$

such that S^{n_0} is an $e^{O(d)}$-approximate group, a normal subgroup $C \lhd G$ of index at most $O_d(1)$ in G, a subgroup $H_0 \lhd C$ with $H_0 \subset S^{\lfloor n/2 \rfloor}$, and $s \ll_d 1$ such that C/H_0 is s-step nilpotent. Writing $C_1 < C_2 < \cdots$ for the lower central series, it follows from Corollary 5.2.9 that $C_{s+1} \subset H_0$. In particular, if we set $H = C_{s+1}$ then by definition of H_0 we have

$$H \subset S^{\lfloor n/2 \rfloor}. \tag{11.4.3}$$

Note also that H is characteristic in C, and hence normal in G by Lemma 1.5.2, and write $\pi : G \to G/H$ for the quotient homomorphism.

Since S^{n_0} is an $e^{O(d)}$-approximate group, there exists a set X of size at most $O_d(1)$ such that $S^{2n_0} \subset XS^{n_0}$. It follows that $S^{mn_0} \subset X^m S^{n_0}$ for every $m \in \mathbb{N}$, and hence that

$$\pi(S^{mn_0}) \subset \pi(X)^m \pi(S^{n_0}) \tag{11.4.4}$$

for every $m \in \mathbb{N}$. We then have

$$\begin{aligned}
|S^{mn_0}| &\le |H||\pi(S^{mn_0})| \\
&\le |H||\pi(X)^m||\pi(S^{n_0})| && \text{(by (11.4.4))} \\
&\le |\pi(X)^m||S^n| && \text{(by (11.4.2), (11.4.3) and Lemma 2.6.3)} \\
&\le |\pi(X)^m|n^d|S| && \text{(by (11.4.1))} \\
&\le |\pi(X)^m|n_0^{2d}|S| && \text{(by (11.4.2))}
\end{aligned}$$

for every $m \in \mathbb{N}$. However, $\pi(X)$ is a subset of G/H of size at most $O_d(1)$, and G/H has an $O_d(1)$-step nilpotent subgroup of index at most $O_d(1)$, namely C/H. Proposition 11.4.3 therefore implies that there exists $k = k_d$ such that $|\pi(X)^m| \le m^k$ for every $m \ge 2$. It follows that whenever $m \ge 2$ and $0 \le r < n_0$ we have

$$\begin{aligned}
|S^{mn_0+r}| &\le ((m+1)n_0)^k|S| \\
&\le (mn_0)^{2k}|S|.
\end{aligned}$$

Since $n > 2n_0$, this implies that $|S^m| \le m^{2k}|S|$ for every $m \ge n$, as required. □

11.5 Diameters of Finite Groups

It is trivial that a finite group has polynomial growth and is virtually nilpotent. Nonetheless, Theorem 11.1.5 and Proposition 11.3.3 still have non-trivial content even when G is finite. When G is finite, a particular value of n for which it is natural to seek to apply Proposition 11.3.3 is the first n for which $S^n = G$. We call this n the *diameter* of G. Indeed, if G is a finite group with a symmetric generating set S containing the identity then the *diameter* $\mathrm{diam}_S(G)$ of G with respect to S is defined via

$$\mathrm{diam}_S(G) = \min\{n \in \mathbb{N} : S^n = G\}.$$

To say that S satisfies the hypothesis (11.3.2) of Proposition 11.3.3 for $n = \mathrm{diam}_S(G)$ is to say that

$$\mathrm{diam}_S(G) \geq \left(\frac{|G|}{|S|}\right)^{1/d}. \tag{11.5.1}$$

For an explicit example of groups satisfying (11.5.1), consider the *discrete torus* T_k^d of side length k and rank d, defined via $T_k^d = (\mathbb{Z}/k\mathbb{Z})^d$. Let $S_{k,d}$ be the generating set for T_k^d consisting of 0 and those elements with one entry equal to ± 1 and all other entries 0. Then it is easy to check that

$$\mathrm{diam}_{S_{k,d}}(T_k^d) \geq d\lfloor \tfrac{1}{2}k \rfloor, \tag{11.5.2}$$

which is proportional to $|T_k^d|^{1/d}$ as $k \to \infty$ for fixed d.

In Exercise 11.7 we invite the reader to show that a finite nilpotent group G also satisfies

$$\mathrm{diam}_S(G) \geq |G|^{1/d}$$

for some d depending only on the step of G and the size of the generating set. Note that some dependence on the generating set is unavoidable here, since an arbitrary finite group G will have diameter 1 if G itself is taken to be the generating set. Nonetheless, writing r for the rank of G, there is always a choice of symmetric generating set containing the identity of size at most $2r + 1$.

It turns out that, just like Gromov's theorem for infinite groups of polynomial growth, there is a converse to Exercise 11.7 stating that every finite group satisfying (11.5.1) is close to nilpotent in some sense. This result is due to Breuillard and the author [20, Theorem 4.1]. We make the statement precise and invite the reader to prove it in Exercise 11.8.

In this section we concentrate on a related result concerning diameters of *finite simple groups*. A group G is said to be *simple* if its only normal subgroups are $\{1\}$ and G. A famous conjecture of Babai (see [1]) asserts that there should exist constants $a, b > 0$ such that if G is an arbitrary non-abelian finite simple group, and if S is a finite symmetric generating set for G containing the identity, then $\text{diam}_S(G) \leq a(\log|A|)^b$.

As we shall describe briefly in the next section, Babai's conjecture is known to hold for certain classes of finite simple groups. However, in general the best bound appears to be the following bound, due to Breuillard and the author.

Corollary 11.5.1 ([20, Corollary 1.4]) *For every $\varepsilon > 0$ there exists $\lambda > 0$ depending only on ε such that if G is a non-abelian finite simple group with a finite symmetric generating set S containing the identity then* $\text{diam}_S(G) \leq \max\{(|G|/|S|)^\varepsilon, \lambda\}$.

Before proving Corollary 11.5.1 we isolate a simple lemma.

Lemma 11.5.2 *Let G be a finite group and let S be a finite symmetric generating set for G containing the identity. Then* $\text{diam}_S(G) < |G|$.

Proof This follows from applying Lemma 11.2.1 with $H = \{1\}$. □

Proof of Corollary 11.5.1 Let λ be a fixed constant depending only on ε, to be determined as the proof progresses. Suppose that $\text{diam}_S(G) > \max\{(|G|/|S|)^\varepsilon, \lambda\}$. Writing $n = \text{diam}_S(G)$, the hypothesis means that $|S^n| < n^{1/\varepsilon}$, so provided λ is large enough we may apply Proposition 11.3.3 to obtain $H \lhd C < G$ with $H \subset S^{\lfloor n/2 \rfloor}$ and $[G : C] \leq O_\varepsilon(1)$ such that C/H is nilpotent of step at most $O_\varepsilon(1)$. If $|G|$ is large enough in terms of ε then Lemma 11.2.3 implies that $C = G$ (as G simple); Lemma 11.5.2 implies that this is the case if λ is large enough. This means that $H \lhd G$, and since $H \subset S^{\lceil n/2 \rceil}$ we have $H \neq G$, so it must be that $H = \{1\}$. Thus G is nilpotent. Corollary 5.2.9 therefore implies that the lower central series for G terminates at $\{1\}$ in finitely many steps, and so Proposition 5.2.4 implies that $[G, G] \neq G$. Since G is simple and $[G, G] \lhd G$, it follows that G is abelian, as required. □

11.6 An Isoperimetric Inequality for Finite Groups

In this section we show how Corollary 11.4.1 can be used to derive a so-called *isoperimetric inequality* for finite groups. The inequality is in itself quite natural, but it is also has links to bounding resistances and studying random walks on vertex-transitive electric networks; for more details and motivation regarding this aspect the reader may consult the paper [4] of Benjamini and Kozma, or the forthcoming paper [70] of Tessera and the author.

In order to state the inequality we are going to prove in this section, we must first introduce some terminology. Given a group G with a finite symmetric generating set S containing the identity, the *boundary* ∂A of a subset $A \subset G$ with respect to S is defined via

$$\partial A = AS \setminus A.$$

An *isoperimetric inequality* for G is a lower bound on $|\partial A|$ in terms of $|A|$. For example, in the discrete torus $T_k^d = (\mathbb{Z}/k\mathbb{Z})^d$ with $S = S_{k,d}$ the standard generating set we have

$$|\partial A| \gg_d |A|^{\frac{d-1}{d}} \tag{11.6.1}$$

for every $A \subset T_k^d$ with $|A| \leq \frac{1}{2}|T_k^d|$. Moreover, it is easy to see that this is the best possible bound, since the set nS satisfies $|\partial(nS)| \ll_d |nS|^{\frac{d-1}{d}}$.

In forthcoming work, Tessera and the author prove that the diameter bound (11.5.2) on the discrete torus is by itself essentially enough to deduce the isoperimetric inequality (11.6.1), as follows. This confirms a conjecture of Benjamini and Kozma [4, Conjecture 4.1].

Theorem 11.6.1 ([70]) *Let $d \geq 1$, and suppose that G is a finite group with a symmetric generating set S containing the identity such that* $\operatorname{diam}_S(G) \leq |G|^{1/d}$. *Then for every subset $A \subset \Gamma$ with $|A| \leq |\Gamma|/2$ we have $|\partial A| \gg_d |A|^{\frac{d-1}{d}}$.*

The proof of Theorem 11.6.1 rests on the version of Corollary 11.4.1 described in Remark 11.4.2, a proof of which is beyond the scope of this book. Nonetheless, we can illustrate the idea by proving the following variant of Theorem 11.6.1.

Corollary 11.6.2 (Breuillard–Green–Tao [18, Corollary 11.15]) *There exist $b_0 \geq 0$ and functions $c, d : [b_0, \infty) \to [0, \infty)$ with $d(b) \to \infty$ as $b \to \infty$ such that, for any $b \geq b_0$, if G is a finite group with a symmetric generating set S containing the identity such that* $\operatorname{diam}_S(G) \leq (|G|/|S|)^{1/b}$ *then for every subset $A \subset G$ with $c(b)|S| \leq |A| \leq |G|/2$ we have*

$$|\partial A| \geq \tfrac{1}{16}|S|^{\frac{1}{d(b)}}|A|^{\frac{d(b)-1}{d(b)}}.$$

Tessera and the author optimise the function d in the forthcoming paper [70], but that is beyond the scope of this book.

The ability to deduce isoperimetric inequalities such as Theorem 11.6.1 and Corollary 11.6.2 from results about growth such as Corollary 11.4.1 comes from the following result.

Proposition 11.6.3 (Coulhon–Saloff-Coste) *Let G be a finite group with a symmetric generating set S containing the identity. Define $\phi : [|G|] \to [\mathrm{diam}_S(G)]$ by setting*

$$\phi(n) = \min\{r \in \mathbb{N} : |S^r| \geq n\}.$$

Then for every $A \subset G$ with $1 \leq |A| \leq |G|/2$ we have $|A| \leq 4\phi(2|A|)|\partial A|$.

Proposition 11.6.3 shows that in order to prove Corollary 11.6.2 it is sufficient to prove that $\phi(n)$ grows 'slowly' as n grows, or equivalently that $|S^r|$ grows 'quickly' as r grows. We achieve this through the following consequence of Corollary 11.4.1.

Corollary 11.6.4 (Breuillard–Green–Tao [18, Corollary 11.10]) *There exist $b_0 \geq 0$ and functions $d, r_0 : [b_0, \infty) \to [0, \infty)$ with $d(b) \to \infty$ as $b \to \infty$ such that, for any $b \geq b_0$, if G is a finite group with a symmetric generating set S containing the identity such that $\mathrm{diam}_S(G) \leq (|G|/|S|)^{1/b}$ then $|S^r| \geq \min\{r^{d(b)}|S|, |G|\}$ for every $r \geq r_0(b)$.*

We start by proving Proposition 11.6.3, following the argument given in [18]. After that we prove Corollary 11.6.4, before finally combining them to prove Corollary 11.6.2.

In proving Proposition 11.6.3 we define a linear action of G on $\ell^1(G)$ via $g \cdot f(x) = f(g^{-1}x)$. Note that this action is an isometry for the ℓ^1 norm $\|\cdot\|_1$, in the sense that

$$\|g \cdot f\|_1 = \|f\|_1 \tag{11.6.2}$$

for every $g \in G$ and $f \in \ell^1(G)$. Given a finite set $B \subset G$, we define the linear operator $M_B : \ell^1(G) \to \ell^1(G)$ via

$$M_B(f) = \mathbb{E}_{b \in B}\, b \cdot f.$$

Lemma 11.6.5 *Let $f \in \ell^1(G)$. Then for every $g \in G$ we have $\|g \cdot f - f\|_1 \leq |g| \max_{s \in S} \|s \cdot f - f\|_1$. In particular, for every $B \subset S^n$ we have $\|M_B(f) - f\|_1 \leq n \max_{s \in S} \|s \cdot f - f\|_1$.*

Proof Writing $g = s_1 \cdots s_r$ with $s_i \in S$ and $r = |g|$, the triangle inequality gives

$$\begin{aligned}
\|g \cdot f - f\|_1 &\leq \sum_{i=1}^{r} \|s_1 \cdots s_i \cdot f - s_1 \cdots s_{i-1} f\|_1 \\
&= \sum_{i=1}^{r} \|s_1 \cdots s_{i-1}(s_i \cdot f - f)\|_1 \\
&= \sum_{i=1}^{r} \|s_i \cdot f - f\|_1 && \text{by (11.6.2)} \\
&\leq r \max_{s \in S} \|s \cdot f - f\|_1,
\end{aligned}$$

as required. □

Lemma 11.6.6 *Let $A \subset G$ be a finite set and let $s \in S$. Then* $|As \,\triangle\, A| \leq 2|\partial A|$.

Proof We have $|As \setminus A| \leq |\partial A|$ by definition of $|\partial A|$. On the other hand, if $x \in A \setminus As$ then $xs^{-1} \in As^{-1} \setminus A \subset \partial A$ by symmetry of S, and so $A \setminus As \subset (\partial A)s$, giving $|A \setminus As| \leq |\partial A|$ as well and completing the proof. □

Proof of Proposition 11.6.3 Following [18, Lemma 11.16], first note that, writing 1_A for the indicator function of A, we have

$$\begin{aligned}
\|g \cdot 1_A - 1_A\|_1 &= |gA \,\triangle\, A| \\
&= |A^{-1}g^{-1} \,\triangle\, A^{-1}|
\end{aligned}$$

for every $g \in G$, and hence in particular

$$\begin{aligned}
\|s^{-1} \cdot 1_{A^{-1}} - 1_{A^{-1}}\|_1 &\leq |As \,\triangle\, A| \\
&\leq 2|\partial A| && \text{(by Lemma 11.6.6).}
\end{aligned}$$

Combined with Lemma 11.6.5, this implies that

$$\|M_{S^r}(1_{A^{-1}}) - 1_{A^{-1}}\|_1 \leq 2r|\partial A| \tag{11.6.3}$$

for every $r \in \mathbb{N}$.

On the other hand, note that for every $f \in \ell^1(G)$ and every finite subset $B \subset G$ we have

$$\|M_B(f)\|_\infty \leq \frac{\|f\|_1}{|B|}.$$

This implies in particular that

$$\|M_{S^r}(1_{A^{-1}})\|_\infty \leq \frac{|A|}{|S^r|}.$$

If $|S^r| \geq 2|A|$, which is to say if $r \geq \phi(2|A|)$ (which is well defined for all A with $1 \leq |A| \leq |G|/2$), this in turn means that $\|M_{S^r}(1_{A^{-1}})\|_\infty \leq \frac{1}{2}$, and hence that

$$\|M_{S^r}(1_{A^{-1}}) - 1_{A^{-1}}\|_1 \geq \frac{|A|}{2}.$$

Applying this in the case $r = \phi(2|A|)$ and combining it with (11.6.3) then proves the proposition. □

Proof of Corollary 11.6.4 For every $b > 0$, let U_b be the set of those $d > 0$ for which the constant implicit in the $O_d(1)$ notation of Corollary 11.4.1 can be taken to be $\frac{1}{2}b$. Fix b_0 so that $\frac{1}{2}b_0$ is a possible constant implicit in that $O_d(1)$ notation when $d = 1$, so that $U_b \neq \varnothing$ for every $b \geq b_0$. Then, for every $b \geq b_0$, set $d(b) = \frac{1}{2}\sup U_b$, and define $r_0(b) \geq 2$ to be the constant $N_{d(b)}$ arising from Corollary 11.4.1, noting that $d(b) \to \infty$ as $b \to \infty$ by Corollary 11.4.1.

Let $b \geq b_0$ and $r \geq r_0(b)$. Suppose that G is a finite group with a symmetric generating set S such that $\mathrm{diam}_S(G) > r$ and $|S^r| < r^{d(b)}|S|$. By definition of d, Corollary 11.4.1 implies that $|S^m| < m^b|S|$ for every $m \geq r$. In the particular case of $m = \mathrm{diam}_S(G)$, this translates to $|G| < \mathrm{diam}_S(G)^b|S|$, and hence $\mathrm{diam}_S(G) > (|G|/|S|)^{1/b}$. This proves the corollary. □

Proof of Corollary 11.6.2 Take b_0, d and r_0 as in Corollary 11.6.4, and let ϕ be defined as in Proposition 11.6.3. Suppose that $b \geq b_0$. Corollary 11.6.4 then implies that for every $n \in \mathbb{N}$ with $r_0(b)^{d(b)}|S| \leq n < |G|$ we have

$$\phi(n) \leq \left\lceil \left(\frac{n}{|S|}\right)^{\frac{1}{d(b)}} \right\rceil \leq 2\left(\frac{n}{|S|}\right)^{\frac{1}{d(b)}}.$$

Combined with Proposition 11.6.3, this implies that for every $A \subset G$ with $\frac{1}{2}r_0(b)^{d(b)}|S| \leq |A| \leq |G|/2$ we have

$$|\partial A| \geq \tfrac{1}{16}|S|^{\frac{1}{d(b)}}|A|^{1-\frac{1}{d(b)}},$$

as required. □

11.A Expansion in Special Linear Groups

In this appendix we briefly discuss applications of approximate groups to *expansion*. Expansion is a property of considerable importance in a number of contexts in pure and applied mathematics, and particularly theoretical computer science. A detailed introduction to expansion is beyond the scope of this book, but there are many excellent books and surveys on expanders to which the reader could turn. For a thorough general introduction to expanders, including a number of examples of applications, the reader could in the first instance consult Hoory, Linial and Wigderson's survey [41]. For a detailed description of the topics discussed in this appendix, see Tao's book [66].

To motivate our discussion, first note that Proposition 11.3.3, and hence Corollary 11.5.1, ultimately rest on Theorem 7.1.1, the bounds in which are not good enough to imply Babai's conjecture. It turns out, however, that there are certain classes of finite simple groups for which much stronger results than Theorem 7.1.1 are known and do lead to Babai-type bounds.

In this appendix we focus on linear groups over finite fields. Given a field $\mathbb{K}$ and $n \in \mathbb{N}$, the *special linear group* $SL_n(\mathbb{K})$ is defined via $SL_n(\mathbb{K}) = \{g \in GL_n(\mathbb{K}) : \det g = 1\}$. The group $SL_n(\mathbb{K})$ is in general not simple, since the centre $Z(SL_n(\mathbb{K}))$ consists of all those scalar transformations contained in $SL_n(\mathbb{K})$. However, on quotienting by this centre we arrive at the *projective special linear group*

$$PSL_n(\mathbb{K}) = SL_n(\mathbb{K})/Z(SL_n(\mathbb{K})),$$

which is a finite simple group whenever $n \geq 2$ and $\mathbb{K}$ is finite (except if $n = 2$ and $\mathbb{K} = \mathbb{F}_2$ or $\mathbb{F}_3$).

In $SL_n(\mathbb{K})$ we have the following theorem, which was announced independently by Breuillard, Green and Tao [15] and Pyber and Szabo [48] within four hours of one another! It followed work of Helfgott [39, 40], who had previously treated the special cases of $SL_2(\mathbb{F}_p)$ and $SL_3(\mathbb{F}_p)$.

Theorem 11.A.1 ([66, Theorem 1.5.1]) *Let $\mathbb{K}$ be a finite field, let $n \geq 2$, and let A be a generating set for $SL_n(\mathbb{K})$. Suppose $\varepsilon > 0$ is small enough depending only on n. Then either $|A^3| \geq |A|^{1+\varepsilon}$ or $|A| \geq |SL_n(\mathbb{K})|^{1-O_n(\varepsilon)}$.*

In Exercise 11.9 we invite the reader to show that Theorem 11.A.1 implies Babai-type diameter bounds for $SL_n(\mathbb{K})$ and $PSL_n(\mathbb{K})$.

In fact, it turns out that Theorem 11.A.1 is also enough to prove the related property of *expansion*. At heart, expansion is really a property of graphs, rather than groups. An *expander graph* is, roughly, a graph that is both sparse and highly connected. More precisely, given a subset A of a graph Γ, define the *edge boundary* $\partial_e A$ of A to be the set of edges between A and $\Gamma \setminus A$. The *expansion ratio* or *Cheeger constant* $h(\Gamma)$ of a finite graph Γ is then defined via

$$h(\Gamma) = \min_{A \subset \Gamma : |A| \leq |\Gamma|/2} \frac{|\partial_e A|}{|A|}.$$

Given $\varepsilon > 0$ and $d \in \mathbb{N}$, a family X of finite graphs is said to be a *family of (d, ε)-expanders*, if $h(\Gamma) \geq \varepsilon$ for every $\Gamma \in X$, if each vertex of each graph in X has degree at most d, and if $\sup_{\Gamma \in X} |\Gamma| = \infty$. We say that X is a *family of expanders* if there exist some $\varepsilon > 0$ and $d \in \mathbb{N}$ such that X is a family of (d, ε)-expanders.

The lower bound on $h(\Gamma)$ for $\Gamma \in X$ provides the sense in which expander families are highly connected. On the other hand, the upper bound on the degrees of the vertices means that the number of edges in a graph $\Gamma \in X$ is at most linear in $|\Gamma|$, which is asymptotically as sparse as one could hope to make a family of connected graphs. The requirement that $\sup_{\Gamma \in X} |\Gamma| = \infty$ exists to prevent every finite family of graphs from trivially being a family of expanders.

To give a simple reason why families of expanders are of interest, note that sparsity and high connectivity are both desirable properties of communication or transport networks, but are intuitively difficult to achieve simultaneously.

One of the difficulties of the theory of expander graphs is to construct families of expanders, and it turns out that such families can be constructed using groups. A very natural way of obtaining a graph from a group is via a construction called a *Cayley graph*. Given a group G with a symmetric generating set S, the *Cayley graph* $\Gamma(G, S)$ of G with respect to S is the graph whose vertices are the elements of G, with an edge between $x \in G$ and $y \in G$ if there exists $s \in S$ such that $x = ys$.

It turns out that, using a remarkable argument developed by Bourgain and Gamburd, Theorem 11.A.1 can be used to prove that certain families of Cayley graphs of $SL_n(\mathbb{Z}/p\mathbb{Z})$ with p prime are expander families. Indeed, Theorem 11.A.1 and this argument show that, given a generating set S of $SL_n(\mathbb{Z})$, if we write S_p for the image of S in $SL_n(\mathbb{Z}/p\mathbb{Z})$ then the Cayley graphs $\Gamma(SL_n(\mathbb{Z}/p\mathbb{Z}), S_p)$ with p prime form a family of expanders. See Green's survey [33] for an excellent sketch of the Bourgain–Gamburd argument; see Tao's book [66] or Pyber and Szabo's paper [49] for fuller details. It is well known (see [20, Lemma 5.1], for example) that this implies in particular the bounds $\operatorname{diam}_{S_p}(SL_n(\mathbb{Z}/p\mathbb{Z})) \ll_{n,S} \log p$.

Exercises

11.1 Show that an infinite finitely generated group has at least linear growth, in the sense that for every infinite group G generated by a finite set S there exists $c > 0$ such that $|S^n| \geq cn$.

11.2 Let F be the free group on generators x, y, and let $S = \{1, x, x^{-1}, y, y^{-1}\}$. Show that $|S^n| = 2 \cdot 3^n - 1$ for every $n \in \mathbb{N}$.

11.3

(a) Show that for every $K \geq 1$ there exists $N = N_K$ such that if G is a group generated by a finite symmetric set S containing the identity, and if A is a finite subset of G satisfying $A \supset S^N$ and $|A^2| \leq K|A|$, then there exists a finite normal subgroup $H \lhd G$ and a normal subgroup $C \lhd G$ of index at most $O_K(1)$ such that C/H is nilpotent of rank and step at most $O_K(1)$.

(b) Show that if A is assumed to be a K-approximate group in part (a) then we may conclude, in addition, that A^4 contains H and a generating set for C.

(c) Verify that the results of each of parts (a) and (b) refine Theorem 11.1.6.

11.4 Show that for every $C > 0$ and $d \geq 0$ there exists $N = N_d$ such that if G is a group generated by a finite symmetric set S containing the identity, and if $|S^n| \leq Cn^d$ for some $n \geq N$, then G has an $O_d(1)$-step nilpotent subgroup of index at most $O_d((Cn^d)!)$.

11.5 Show that there exists an absolute constant $c > 0$ such that if G is a residually nilpotent group with finite symmetric generating set S containing the identity, and if there exists $n > 1$ such that $|S^n| \leq n^{c \log \log n}$, then G contains a $(\log n)$-step nilpotent subgroup of index $O_n(1)$.

11.6 Show that for each $d \in \mathbb{N}$ there exists a constant $c = c_d > 0$ such that if G is a soluble subgroup of $GL_d(\mathbb{C})$ with a finite symmetric generating set S containing the identity and satisfying $|S^n| \leq n^{c \log n}|S|$ for some $n \geq 2$, then G is virtually d-step nilpotent.

11.7 Let G be a finite s-step nilpotent group generated by a symmetric subset S containing the identity. Show that there exists $\varepsilon = \varepsilon(s, |S|) > 0$ depending only on s and $|S|$ such that $\mathrm{diam}_S(G) \geq |G|^\varepsilon$. *Hint: Show first that if G is a nilpotent group of rank r and step s then $|G/[G,G]| \geq |G|^{\Omega_{r,s}(1)}$.*

11.8 Let G be a finite group generated by a symmetric subset S containing the identity such that

$$\mathrm{diam}_S(G) \geq \left(\frac{|G|}{|S|}\right)^{1/d}.$$

Show that, provided $\mathrm{diam}_S(G)$ is large enough in terms of ε only, the following hold.

(a) G has a normal subgroup H contained in $S^{\lfloor \mathrm{diam}_S(G)^{1/2} \rfloor}$ such that G/H has a nilpotent subgroup of index, rank and step at most $O_\varepsilon(1)$.

(b) G has a normal subgroup H contained in $S^{\lfloor \mathrm{diam}_S(G)^{3/4} \rfloor}$ such that G/H has an abelian subgroup of index and rank at most $O_\varepsilon(1)$.

11.9 Deduce from Theorem 11.A.1 that if $\mathbb{K}$ is a finite field and S is a symmetric generating set containing the identity in $SL_n(\mathbb{K})$ then $\mathrm{diam}_S(SL_n(\mathbb{K})) \ll_n \log^{O_d(1)} |\mathbb{K}|$. Conclude that such a statement also holds with $SL_n(\mathbb{K})$ replaced by $PSL_n(\mathbb{K})$.

11.10 Let G be a group with a symmetric generating set S containing the identity, let H be a subgroup of G, and let T be a subset of G containing the identity such that $HT = G$. Show that the set

$$Y = \{h \in H : \exists s \in S, t \in T, u \in T \text{ such that } h = tsu^{-1})\}$$

generates H. Deduce that Proposition 11.2.2 can be strengthened to say that if H has index $k \in \mathbb{N}$ in G then $S^{2k+1} \cap H$ generates H.

11.11 Suppose that G is a finitely generated group of rank at most r, and that $H \lhd G$ is a normal subgroup of size at most k such that G/H is s-step nilpotent. Show that G has an s-step nilpotent subgroup of index at most $O_{r,k,s}(1)$. *This is a partial refinement of Lemma 11.3.4. Hint: Use Exercise 5.2.*

References

[1] L. Babai and A. Seress, On the diameter of Cayley graphs of the symmetric group. *J. Combin. Theory Ser. A*, **49** (1988), 175–179.

[2] A. Balog and E. Szemerédi, A statistical theorem of set addition. *Combinatorica*, **14**(3) (1994), 263–268.

[3] H. Bass, The degree of polynomial growth of finitely generated nilpotent groups. *Proc. Lond. Math. Soc. Ser. 3*, **25**(4) (1972), 603–614.

[4] I. Benjamini and G. Kozma, A resistance bound via an isoperimetric inequality. *Combinatorica*, **25**(6) (2005), 645–650.

[5] P. Billingsley, *Probability and Measure*, 3rd edn (New York: John Wiley & Sons, 1995).

[6] M. Björklund and T. Hartnick, Approximate lattices. Preprint, arXiv:1612.09246.

[7] H. F. Blichfeldt, A new principle in the geometry of numbers with some applications. *Trans. Amer. Math. Soc.*, **15**(3) (1914), 227–235.

[8] N. N. Bogolyubov, Sur quelques propriétés arithmétiques des presque-périodes. *Ann. Chaire Math. Phys. Kiev*, **4** (1939), 185–194.

[9] J. Bourgain and A. Gamburd, Uniform expansion bounds for Cayley graphs of $SL_2(\mathbb{F}_p)$. *Ann. Math.*, **167**(2) (2008), 625–642.

[10] J. Bourgain, N. Katz and T. Tao, A sum-product estimate in finite fields, and applications. *Geom. Funct. Anal.*, **14**(1) (2004), 27–57.

[11] E. Breuillard, A strong Tits alternative. Preprint, arXiv:0804.1395.

[12] E. Breuillard and B. J. Green, Approximate groups I: The torsion-free nilpotent case. *J. Inst. Math. Jussieu*, **10**(1) (2011), 37–57.

[13] E. Breuillard and B. J. Green, Approximate groups II: The solvable linear case. *Q. J. Math.*, **62**(3) (2011), 513–521.

[14] E. Breuillard and B. J. Green, Approximate groups III: The unitary case. *Turkish J. Math.*, **36**(2) (2012), 199–215.

[15] E. Breuillard, B. J. Green and T. Tao, Linear approximate groups (research announcement). *Electron. Res. Announc. Math. Sci.*, **17** (2010), 57–67.

[16] E. Breuillard, B. J. Green and T. Tao, Approximate subgroups of linear groups. *Geom. Funct. Anal.*, **21**(4) (2011), 774–819.

[17] E. Breuillard, B. J. Green and T. Tao, A note on approximate subgroups of $GL_n(\mathbb{C})$ and uniformly nonamenable groups. Preprint, arXiv:1101.2552.

[18] E. Breuillard, B. J. Green and T. Tao, The structure of approximate groups. *Publ. Math. IHES*, **116**(1) (2012), 115–221.

[19] E. Breuillard, B. J. Green and T. Tao, A nilpotent Freiman dimension lemma. *Eur. J. Combin.*, **34**(8) (2013), 1287–1292.

[20] E. Breuillard and M. C. H. Tointon, Nilprogressions and groups with moderate growth. *Adv. Math.*, **289** (2016), 1008–1055.

[21] J. W. S. Cassels, *An Introduction to the Geometry of Numbers*, corrected reprint of the 1971 edition, Classics in Mathematics (Berlin: Springer-Verlag, 1997).

[22] M. C. Chang, A polynomial bound in Freiman's theorem. *Duke Math. J.*, **113**(3) (2002), 399–419.

[23] E. Croot and O. Sisask, A probabilistic technique for finding almost-periods of convolutions. *Geom. Funct. Anal.*, **20**(6) (2010), 1367–1396.

[24] K. Cwalina and T. Schoen, A linear bound on the dimension in Green–Ruzsa's theorem. *J. Number Theory*, **133** (2013), 1262–1269.

[25] P. de la Harpe, *Topics in Geometric Group Theory*, Chicago Lectures in Mathematics (Chicago: University of Chicago Press, 2000).

[26] G. A. Freiman, *Foundations of a Structural Theory of Set Addition*, Translations of Mathematical Monographs, vol. 37 (Providence, RI: Amer. Math. Soc., 1973). Translated from the 1966 Russian version published by Kazan Gos. Ped. Inst.

[27] G. Freiman, Number-theoretic studies in the Markov spectrum and in the structural theory of set addition. In *Groups and the Inverse Problems of Additive Number Theory* (in Russian) (Moscow: Kalinin. Gos. Univ., 1973).

[28] N. Gill and H. A. Helfgott, Growth in solvable subgroups of $GL_r(\mathbb{Z}/p\mathbb{Z})$. *Math. Ann.*, **360**(1) (2014), 157–208.

[29] A. M. Gleason, Compact subgroups. *Proc. Natl. Acad. Sci. USA*, **37**(9) (1951), 622–623.

[30] W. T. Gowers, A new proof of Szemerédi's theorem for arithmetic progressions of length four. *Geom. Funct. Anal.*, **8**(3) (1998), 529–551.

[31] W. T. Gowers, A new proof of Szemerédi's theorem. *Geom. Funct. Anal.*, **11**(3) (2001), 465–588.

[32] W. T. Gowers, A new way of proving sumset estimates. Available at https://gowers.wordpress.com/2011/02/10/a-new-way-of-proving-sumset-estimates/.

[33] B. J. Green, Approximate groups and their applications: Work of Bourgain, Gamburd, Helfgott and Sarnak, arXiv:0911.3354.

[34] B. J. Green and I. Z. Ruzsa, Sets with small sumset and rectification. *Bull. Lond. Math. Soc.*, **38**(1) (2006), 43–52.

[35] B. J. Green and I. Z. Ruzsa, Freiman's theorem in an arbitrary abelian group. *J. Lond. Math. Soc.*, **75**(1) (2007), 163–175.

[36] M. Gromov, Groups of polynomial growth and expanding maps. *Publ. Math. IHES*, **53** (1981), 53–73.

[37] R. M. Guralnick, On a result of Schur. *J. Algebra*, **59** (1979), 302–310.

[38] M. Hall, *The Theory of Groups* (Providence, RI: Amer. Math. Soc./Chelsea, 1999).

[39] H. A. Helfgott, Growth and generation in $SL_2(\mathbb{Z}/p\mathbb{Z})$. *Ann. Math.*, **167**(2) (2008), 601–623.

[40] H. A. Helfgott, Growth in $SL_3(\mathbb{Z}/p\mathbb{Z})$. *J. Eur. Math. Soc.*, **13**(3) (2011), 761–851.

[41] S. Hoory, N. Linial and A. Wigderson, Expander graphs and their applications. *Bull. Amer. Math. Soc.*, **43** (2006), 439–561.

[42] E. Hrushovski, Stable group theory and approximate subgroups. *J. Amer. Math. Soc.*, **25**(1) (2012), 189–243.

[43] S. Lovett and O. Regev. A counterexample to a strong variant of the polynomial Freiman-Ruzsa conjecture, *Discrete Anal.* (2017), article no. 8.

[44] A. I. Mal'cev, On certain classes of infinite soluble groups. *Mat. Sb.*, **28** (1951), 567–588 (in Russian). *Amer. Math. Soc. Transl. Ser. 2*, **45** (1956), 1–21.

[45] G. Petridis, New proofs of Plünnecke-type estimates for product sets in groups. *Combinatorica*, **32**(6) (2012), 721–733.

[46] G. Petridis, Upper bounds on the cardinality of higher sumsets. *Acta Arith.*, **158**(4) (2013), 299–319.

[47] H. Plünnecke, Eine zahlentheoretische Anwendung der Graphentheorie. *J. Reine Angew. Math.*, **243** (1970), 171–183.

[48] L. Pyber and E. Szabó, Growth in finite simple groups of Lie type. Research announcement, arXiv:1001.4556.

[49] L. Pyber and E. Szabó, Growth in finite simple groups of Lie type. *J. Amer. Math. Soc.*, **29**(1) (2016), 95–146.

[50] A. A. Razborov, A product theorem in free groups. *Ann. Math.*, **179**(2) (2014), 405–429.

[51] D. J. S. Robinson, *A Course in the Theory of Groups*, 2nd edn, Graduate Texts in Mathematics (New York: Springer-Verlag, 1995).

[52] I. Z. Ruzsa, On the cardinality of $A + A$ and $A - A$. In *Combinatorics*, Coll. Math. Soc. J. Bolyai, vol. 18, (Budapest: North-Holland / Bolyai Társulat, Budapest, 1978), 933–938.

[53] I. Z. Ruzsa, An application of graph theory to additive number theory. *Scientia, Ser. A*, **3** (1989), 97–109.

[54] I. Z. Ruzsa, Addendum to: An application of graph theory to additive number theory. *Scientia, Ser. A*, **4** (1990/91), 93–94.

[55] I. Z. Ruzsa, Generalized arithmetical progressions and sumsets. *Acta Math. Hungar.*, **65**(4) (1994), 379–388.

[56] I. Z. Ruzsa, An analog of Freiman's theorem in groups. *Astérisque*, **258** (1999), 323–326.

[57] S. R. Safin, Powers of subsets of free groups. *Mat. Sb.*, **202**(11) (2011), 97–102. English translation in *Sb. Math.*, **202**(11–12) (2011), 1661–1666.

[58] T. Sanders, On a non-abelian Balog–Szemerédi-type lemma. *J. Aust. Math. Soc.*, **89**(1) (2010), 127–132.

[59] T. Sanders, On the Bogolyubov–Ruzsa lemma. *Anal. PDE*, **5**(3) (2012), 627–655.

[60] J. Solymosi, On sum-sets and product-sets of complex numbers. *J. Théor. Nombres Bordeaux*, **17** (2005), 921–924.
[61] E. Szemerédi, On sets of integers containing no k elements in arithmetic progression. *Acta Arith.*, **27** (1975), 199–245.
[62] T. Tao, Product set estimates for non-commutative groups. *Combinatorica*, **28**(5) (2008), 547–594.
[63] T. Tao, The sum-product phenomenon in arbitrary rings. *Contrib. Discrete Math.*, **4**(2) (2009), 59–82.
[64] T. Tao, An elementary non-commutative Freiman theorem. Available at https://terrytao.wordpress.com/2009/11/10/an-elementary-non-commutative-freiman-theorem/
[65] T. Tao, Freiman's theorem for solvable groups. *Contrib. Discrete Math.*, **5**(2) (2010), 137–184.
[66] T. Tao, *Expansion in Finite Simple Groups of Lie Type*, Graduate Studies in Mathematics, vol. 164 (Providence, RI: Amer. Math. Soc., 2015).
[67] T. Tao, Inverse theorems for sets and measures of polynomial growth. *Q. J. Math.*, **68**(1) (2017), 13–57.
[68] T. Tao and V. H. Vu, *Additive Combinatorics*, Cambridge Studies in Advanced Mathematics, vol. 105 (Cambridge: Cambridge University Press, 2006).
[69] R. Tessera and M. C. H. Tointon, Properness of nilprogressions and the persistence of polynomial growth of given degree. *Discrete Anal.* (2018), article no. 17.
[70] R. Tessera and M. C. H. Tointon, Sharp relations between volume growth, isoperimetry and resistance in vertex-transitive graphs. In preparation.
[71] J. Tits, Free subgroups in linear groups. *J. Algebra*, **20** (1972), 250–270.
[72] M. C. H. Tointon, Freiman's theorem in an arbitrary nilpotent group. *Proc. Lond. Math. Soc. Ser. 3*, **109** (2014), 318–352.
[73] M. C. H. Tointon, Approximate subgroups of residually nilpotent groups. *Math. Ann.*, **374** (2019), 499–515.
[74] M. C. H. Tointon, Polylogarithmic bounds in the nilpotent Freiman theorem. Preprint, arXiv:1812.06735. To appear in *Math. Proc. Camb. Phil. Soc.*
[75] B. A. F. Wehrfritz, *Infinite Linear Groups* (Berlin: Springer-Verlag, 1973).

Index